Optimizing U.S. Air Force and Department of Defense Review of Air Force Acquisition Programs

Committee on Optimizing U.S. Air Force and Department of Defense Review of Air Force Acquisition Programs

Air Force Studies Board

Division on Engineering and Physical Sciences

NATIONAL RESEARCH COUNCIL
OF THE NATIONAL ACADEMIES

THE NATIONAL ACADEMIES PRESS
Washington, D.C.
www.nap.edu

THE NATIONAL ACADEMIES PRESS 500 Fifth Street, N.W. Washington, DC 20001

NOTICE: The project that is the subject of this report was approved by the Governing Board of the National Research Council, whose members are drawn from the councils of the National Academy of Sciences, the National Academy of Engineering, and the Institute of Medicine. The members of the committee responsible for the report were chosen for their special competences and with regard for appropriate balance.

This is a report of work supported by Grant FA9550-08-1-0029 between the U.S. Air Force and the National Academy of Sciences. Any opinions, findings, conclusions, or recommendations expressed in this publication are those of the author(s) and do not necessarily reflect the view of the organizations or agencies that provided support for the project.

International Standard Book Number-13: 978-0-309-13918-2
International Standard Book Number-10: 0-309-13918-X

Limited copies of this report are available from:

Air Force Studies Board
National Research Council
500 Fifth Street, N.W.
Washington, DC 20001
(202) 334-3111

Additional copies are available from:

The National Academies Press
500 Fifth Street, N.W.
Lockbox 285
Washington, DC 20055
(800) 624-6242 or (202) 334-3313
(in the Washington metropolitan area)
Internet, http://www.nap.edu

Printed in the United States of America

THE NATIONAL ACADEMIES

Advisers to the Nation on Science, Engineering, and Medicine

The **National Academy of Sciences** is a private, nonprofit, self-perpetuating society of distinguished scholars engaged in scientific and engineering research, dedicated to the furtherance of science and technology and to their use for the general welfare. Upon the authority of the charter granted to it by the Congress in 1863, the Academy has a mandate that requires it to advise the federal government on scientific and technical matters. Dr. Ralph J. Cicerone is president of the National Academy of Sciences.

The **National Academy of Engineering** was established in 1964, under the charter of the National Academy of Sciences, as a parallel organization of outstanding engineers. It is autonomous in its administration and in the selection of its members, sharing with the National Academy of Sciences the responsibility for advising the federal government. The National Academy of Engineering also sponsors engineering programs aimed at meeting national needs, encourages education and research, and recognizes the superior achievements of engineers. Dr. Charles M. Vest is president of the National Academy of Engineering.

The **Institute of Medicine** was established in 1970 by the National Academy of Sciences to secure the services of eminent members of appropriate professions in the examination of policy matters pertaining to the health of the public. The Institute acts under the responsibility given to the National Academy of Sciences by its congressional charter to be an adviser to the federal government and, upon its own initiative, to identify issues of medical care, research, and education. Dr. Harvey V. Fineberg is president of the Institute of Medicine.

The **National Research Council** was organized by the National Academy of Sciences in 1916 to associate the broad community of science and technology with the Academy's purposes of furthering knowledge and advising the federal government. Functioning in accordance with general policies determined by the Academy, the Council has become the principal operating agency of both the National Academy of Sciences and the National Academy of Engineering in providing services to the government, the public, and the scientific and engineering communities. The Council is administered jointly by both Academies and the Institute of Medicine. Dr. Ralph J. Cicerone and Dr. Charles M. Vest are chair and vice chair, respectively, of the National Research Council.

www.national-academies.org

COMMITTEE ON OPTIMIZING U.S. AIR FORCE AND DEPARTMENT OF DEFENSE REVIEW OF AIR FORCE ACQUISITION PROGRAMS

RAND H. FISHER, *Chair*, The Aerospace Corporation
J. DANIEL STEWART, *Vice Chair,* University of Tennessee
JOHN A. BETTI, Department of Defense (retired)
CHRISTOPHER L. BLAKE, Lockheed Martin Aeronautics
CLAUDE M. BOLTON, JR., Defense Acquisition University
ALLAN V. BURMAN, Jefferson Solutions
JOHN T. DILLARD, U.S. Naval Postgraduate School
CHARLES E. FRANKLIN, Raytheon Company (retired)
CHARLES L. JOHNSON II, Boeing Company
LESLIE F. KENNE, LK Associates
ANDREW P. SAGE, George Mason University
MARK D. SCHAEFFER, ManTech SRS Technologies
GEORGE R. SCHNEITER, Consultant
ROBERT J. SKALAMERA, RJS Consulting
RICHARD SZAFRANSKI, Toffler Associates
RANDALL S. WEIDENHEIMER, Northrop Grumman Mission Systems
REBECCA A.WINSTON, Winston Strategic Management Consultants

Staff

JAMES C. GARCIA, Study Director
KAMARA E. BROWN, Research Associate
ZEIDA PATMON, Program Associate
MARGUERITE SCHNEIDER, Administrative Coordinator
ENITA A. WILLIAMS, Research Associate (Until October 2008)
URRIKKA B. WOODS, Program Associate

AIR FORCE STUDIES BOARD

ZEIDA PATMON, Program Associate
MARGUERITE SCHNEIDER, Administrative Coordinator
DANIEL E.J. TALMAGE, JR., Program Officer
SHANNON C. THOMAS, Program Associate
ENITA A. WILLIAMS, Research Associate (until October 2008)
URRIKKA B. WOODS, Program Associate

Preface

The Deputy Assistant Secretary of the Air Force for Science, Technology, and Engineering asked the National Research Council (NRC) to review and assess the increasing number of prescribed program reviews and assessments that U.S. Air Force space and nonspace acquisition programs in all Department of Defense (DOD) acquisition categories (ACATs) are required to undergo and to recommend ways to improve the effectiveness and efficiency of those program reviews in terms of their goals, objectives, content, and requirements. The Committee on Optimizing U.S. Air Force and DOD Review of Air Force Acquisition Programs was formed in May 2008 to conduct this review.

In developing its approach to the study, the committee concluded that the most important objective of the program reviews referred to above is to contribute to the successful execution of acquisition programs. To this end, the committee examined the substantial library of past studies; conducted a variety of interviews; surveyed managers of acquisition programs and program executive officers (PEOs); and constructed a matrix to summarize program reviews in terms of their purpose, target audience, and product output. Additionally, the committee is aware that substantial changes in acquisition policies and their implementation are being considered.

The committee's findings, conclusions, and recommendations address principally the subject of program reviews and also make some observations on other critical elements of acquisition success.

The committee acknowledges and appreciates the contribution of the members of the Air Force Studies Board (AFSB) of the NRC for developing the study statement of task in concert with the Air Force sponsor. The AFSB was established in 1996 as a unit of the NRC at the request of the U.S. Air Force. The AFSB brings to bear broad military, industrial, and academic scientific, engineering, and management expertise on Air Force technical challenges and other issues of importance to senior Air Force leaders. It discusses potential studies of interest, develops and frames study tasks, ensures proper project planning, suggests potential committee members and reviewers for reports produced by fully independent ad hoc study committees, and convenes meetings to examine strategic issues. The board members, listed on page *v*, were not asked to endorse the committee's conclusions or recommendations, nor did they review the final draft of this report before its release, although board members with appropriate expertise may be nominated to serve as formal members of study committees (as were Rand Fisher, Dan Stewart, and Leslie Kenne in this case) or as report reviewers.

The committee thanks the many persons who helped provide information to the committee, including all the guest speakers shown in Appendix B, their organizations, and supporting staff members; all the Air Force program managers and program executive officers who completed the committee's survey; the Air Force sponsor of this study, Terry Jaggers, and his staff members, including Lt Col Ed Masterson and Lt Col Don Hill; and Kristen Baldwin in the Office of the Secretary of Defense. The committee is very grateful to Connie Citro, director of the NRC Committee on National Statistics, for her counsel in developing the committee's survey of Air Force program managers and in evaluating its results. The committee is also grateful to the NRC staff members who provided support throughout the study.

Finally, as chair and vice chair of the study committee, we extend special thanks to the committee members for their commitment and diligence, which enabled us to complete the task successfully.

Rand H. Fisher, *Chair*
J. Daniel Stewart, *Vice Chair*
Committee on Optimizing U.S. Air Force and Department of Defense Review of Air Force Acquisition Programs

Acknowledgment of Reviewers

This report has been reviewed in draft form by individuals chosen for their diverse perspectives and technical expertise, in accordance with procedures approved by the National Research Council's Report Review Committee. The purpose of this independent review is to provide candid and critical comments that will assist the institution in making its published report as sound as possible and to ensure that the report meets institutional standards for objectivity, evidence, and responsiveness to the study charge. The review comments and draft manuscript remain confidential to protect the integrity of the deliberative process. We wish to thank the following individuals for their review of this report:

Edward C. Aldridge, Jr., NAE, U.S. Department of Defense (retired),
Meyer J. Benzakein, NAE, Ohio State University,
Andrew Brown, Jr., NAE, Delphi Corporation,
David E. Crow, NAE, University of Connecticut,
Susan E. Dong, BAE Systems,
James H. Frey, Frey Associates,
Peter Hantos, Aerospace Corporation,
Donald A. Lamontagne, Star Mountain Consulting, Inc.,
Robert H. Latiff, Science Applications International Corporation, and
Robert R. Shannon, NAE, University of Arizona.

Although the reviewers listed above provided many constructive comments and suggestions, they were not asked to endorse the conclusions or recommendations, nor did they see the final draft of the report before its release. The review of this report was overseen by Robert A. Frosch, NAE, Harvard University, and G. Brian Estes, U.S. Navy (retired). Appointed by the National Research Council, they were responsible for making certain that an independent examination of this report was carried out in accordance with institutional procedures and that all review comments were carefully considered. Responsibility for the final content of this report rests entirely with the authoring committee and the institution.

Contents

SUMMARY 1

1 INTRODUCTION 5
Statement of Task, 5
Background, 5
Methodology, 13
Presentations, 13
Literature Review, 14
Survey and Other Data, 14
Comparative Matrix, 14
Integration and Synthesis of Data, 15

2 FINDINGS AND CONCLUSIONS 16
Reviews Have Both Benefits and Costs, 23
Execution of Reviews (Sequencing, Timing, Participation), 24
Program Review Planning, 27

3 RECOMMENDATIONS 30
Implementing the Recommendations, 35

APPENDIXES

A Biographical Sketches of Committee Members 41
B Meetings and Speakers 51
C Related Studies 55
D Survey 58
E Types of Reviews 88

Acronyms

A5	Headquarters Air Force Deputy Chief of Staff for Air, Space, Information Operations, Plans and Requirements (A3/5)
A5R	Headquarters Air Force Deputy Chief of Staff for Air, Space, Information Operations, Plans and Requirements (A3/5), Director of Operational Capability Requirements
AAC	Air Armament Center
ACAT	acquisition category
ACE	acquisition center of excellence
ADM	acquisition decision memorandum
AF	Air Force
AFAA	Air Force Audit Agency
AFCSE	Air Force Center for Systems Engineering
AFMC	Air Force Materiel Command
AFRB	Air Force Review Board
AFROCC	Air Force Requirements for Operational Capability Council
AFSB	Air Force Studies Board
AFSO21	Air Force Smart Operations for the 21st Century
AOTR	assessment of operational test readiness
AP	acquisition plan
APB	acquisition program baseline
ASC	Aeronautical Systems Center
ASD (NII)	Assistant Secretary of Defense for Networks and Information Integration
ASP	acquisition strategy panel
ASR	alternative system review

CD	concept decision
CDR	critical design review
CPI	continuous process improvement
CSAF	Chief of Staff of the Air Force
CSB	configuration steering board
CSP	cost, schedule, performance
DAB	Defense Acquisition Board
DAE	Defense Acquisition Executive
DAES	Defense Acquisition Executive Summary
DAPS	Defense Acquisition Program Support
DASD (NII)	Deputy Assistant Secretary of Defense for Networks and Information Integration
DAU	Defense Acquisition University
DDR&E	Director of Defense Research and Engineering
DOD	Department of Defense
DOD IG	Department of Defense Inspector General
DOT&E	Director of Operational Test and Evaluation
DRR	design readiness review
DSAB	Defense Space Acquisition Board
D&SWS	develop and sustain warfighting systems
DT&E	development, test, and evaluation
DUSD (A&T)	Deputy Under Secretary of Defense for Acquisition and Technology
EDT	Eastern daylight time
ESC	Electronic Systems Center
FCA	functional configuration audit
FOA	field operating agency
FRP	full rate production
GAO	Government Accountability Office
HQ	headquarters
IBR	integrated baseline review
IIPT	integrating integrated product team
IPA	independent program assessment
IPT	integrated product team
J-8	Joint Chiefs of Staff Force Structure, Resources, and Assessment Directorate

JAT	joint assessment team
JCIDS	Joint Capabilities Integration and Development System
JCS	Joint Chiefs of Staff
JROC	Joint Requirements Oversight Council
JROCM	JROC memorandum
KDP	key decision point
KP	knowledge point
KTR	contractor
LHA	logistics health assessment
MAJCOM	major command
MDA	milestone decision authority
MILSATCOM	military satellite communications
MRA	manufacturing readiness assessment
MRL	manufacturing readiness level
MS	milestone
NAS	National Academy of Sciences
NII/AT&L	Networks and Information Integration/Acquisition, Technology, and Logistics
NRC	National Research Council
OC2	oversight, command, and control
OIPT	overarching integrated product team
OPR	office of primary responsibility
OSD	Office of the Secretary of Defense
OT	operational test
OT&E	operational test and evaluation
OTRR	operational test readiness review
PCA	physical configuration audit
PCDRA	post-CDR assessment
PDR	preliminary design review
PEO	program executive officer
PEO/SR	program executive officer sufficiency review
Perf	performance
PM	program manager
PO	Program Office
PRR	production readiness review
PSA	Office of Portfolio Systems Acquisition
PSR	program support review

RADM	rear admiral
RFP	request for proposals
SAE	Service Acquisition Executive
SDD	system development and demonstration
SDO	Services Designated Official
SDR	system design review
SE	system engineering
SEAM	systems engineering assessment model
SECAF	Secretary of the Air Force
SecDef	Secretary of Defense
SES	Senior Executive Service
SFR	system functional review
SMC	Space and Missile Systems Center
SME	subject matter expert
SRR	system requirements review
SSE	Software and Systems Engineering
S&T	Science and Technology
ST&E	Science, Technology, and Engineering
SVR	system verification review
TD	technology development
TRA	technology readiness assessment
TRR	test readiness review
TSAT	Transformational Satellite Communications
USAF	United States Air Force
USD (AT&L)	Under Secretary of Defense for Acquisition, Technology, and Logistics
VCJCS	Vice Chairman, Joint Chiefs of Staff

Summary

The Department of Defense (DOD) spends over $300 billion per year to develop, produce, field, and sustain weapons systems.[1] Achieving success for DOD acquisition programs in an increasingly complex and dynamic arena that spans multiple organizations (including industry) and functions that do not easily align is a challenge. Too often, DOD weapons systems programs experience large cost overruns and schedule delays, contributing to a growing loss of confidence in the DOD acquisition system.[2,3] In response, there has been a growing array of program and technical reviews that a program manager (PM) must face.

While they are one of the essential elements of program success, these reviews result in costs to the program in terms of time spent supporting the reviews at the expense of time lost focusing on program execution. This study addresses a key question: Can changes in the number, content, sequence, or conduct of program reviews help the PM more successfully execute the program?

---

[1] See DOD (U.S. Department of Defense), *National Defense Budget Estimates for FY 2009, Updated September 2008*. This amount is the sum of the amounts shown for "Operation & Maintenance," "Procurement," and "RDT&E." Available online at http://www.defenselink.mil/comptroller/defbudget/fy2009/FY09Greenbook/greenbook_2009_updated.pdf. Last accessed May 19, 2009.

[2] Elizabeth Newell, "GAO: Weapons systems over budget, overdue, underperforming" (April 1, 2008). Available online at http://www.govexec.com/dailyfed/0408/040108e1.htm. Last accessed May 19, 2009.

[3] Government Accountability Office (GAO), *Defense Acquisition: Assessment of Selected Weapons Programs*, GAO-08-467SP, Washington, D.C.: GAO (2008).

METHODOLOGY

The committee was tasked by the Air Force to review the prescribed program management and technical reviews and assessments that U.S. Air Force space and nonspace system acquisition programs are required to undergo; identify and evaluate options for streamlining, tailoring, integrating, or consolidating reviews of programs to increase their cost-effectiveness and to lessen the impact of the reviews on the workforce; and recommend changes that the Air Force and DOD should make. To accomplish its assignment, the committee received presentations by PMs and program executive officers (PEOs) from the three military departments, industry representatives, overseers, practitioners, process owners, and policy writers in DOD, as well as Government Accountability Office (GAO) researchers and others who have studied DOD acquisition in a broader context. The committee studied the pertinent literature on various acquisition reform initiatives in the Air Force, DOD, and other agencies over the last 20 years (see Appendix C). It found very little quantitative information to address all the elements in the Statement of Task. As a result, the committee surveyed Air Force PMs and PEOs to collect quantitative and qualitative information on the impact of external reviews on program execution and to get an idea of how the reviews help them to manage their programs. The committee also gathered information from individual programs on the number and levels of reviews being conducted as part of the current acquisition process. It constructed a comparative matrix to help identify the number and types of known reviews, their purpose and target audiences, all of which could suggest opportunities for streamlining, integrating, and/or consolidating the reviews. The committee deliberated on the results of these efforts and reached a consensus on its findings, conclusions, and recommendations.

As this report was being finished, the Office of the Secretary of Defense (OSD) published revised DOD Instruction 5000.02. The committee's findings, conclusions, and recommendations are not impacted by this revision, which increased the number of program reviews.

RECOMMENDATIONS

The committee presents the following five recommendations aimed at achieving more effective program acquisition and reducing the burden on the PMs. It believes that if the Air Force were to adopt and implement all of the recommendations, it would achieve a "gold standard" that could serve as a benchmark for other DOD acquisition program review efforts.

Recommendation 1. To ensure that they possess a common understanding of the intent, scope, and output of reviews, the Air Force acquisition and requirement communities at all levels should engage in timely planning for program reviews that results in clear, comprehensive, measurable objectives.

To carry out this recommendation, the Service Acquisition Executive (SAE) should direct a governance process that plans, coordinates, and executes reviews at each level of organization. Each program review's objectives, metrics, and success criteria should be effectively communicated to the PM and the office of primary responsibility (OPR) for the review well in advance of the review. To complete the process, a report should be issued by the chair of the review team followed by a closeout report by the PM.

Recommendation 2. The SAE should develop a plan for the timely, synchronized execution of all program reviews. The plan should align with program decision milestones and decision points.

Program reviews should be aligned with program decision points and milestones to ensure that the number of reviews preceding the decision point and milestone reviews is minimized and that the reviews bring value to the program. Properly timed reviews should result in fewer long-term schedule delays and costs, because early identification of issues and risks should allow the PM to institute strategies for managing them. The elimination of some reviews and the combining of others will reduce costs and lighten the burden on the PM and his or her staff.

Recommendation 3. Before creating or approving a new review, the SAE should compare its objectives with those of existing reviews to determine whether one of the latter could accomplish or incorporate those objectives.

This comparison should determine whether broadening the stakeholders for a given review, rather than conducting additional program reviews, would accomplish the objective.

Recommendation 4. The OPR should staff the review team with recognized subject matter experts.

Subject matter experts need to be identified. They should participate in the program review for the review to be of full value to the program. To facilitate this process, the OPR should maintain a roster of subject matter experts in standard technical areas.

Recommendation 5. The OPR conducting the review should ensure that all review outputs are documented, including root causes if any have been identified, and provide recommendations that can be acted upon by the PM, the program management office, or other program stakeholders.

The committee notes that it is a best practice to capture lessons learned, identifying the root cause of problems and risks encountered in program management, as well as to document findings, observations, and recommendations made during a program review. Documenting the output of reviews stimulates open communication and builds an atmosphere of trust that will lead to participation in future program reviews. Another benefit that should result from documenting the output of the review process is that the management and execution of programs will continue to improve, as program personnel are, in effect, mentored by the expertise of the review team.

Although there may not be sufficient data to permit a quantitative response to the question whether changes in the number, content, sequence, or conduct of program reviews can help the PM execute the program more successfully, the committee is confident that if the above recommendations are implemented and rigorously managed by the SAE and his or her staff, there will be greater control of the review process, which will directly benefit the PMs and allow the successful execution of their programs. The continual learning process that these recommendations represent exemplifies a program management learning process that builds from one review to the next.

1

Introduction

STATEMENT OF TASK

The committee was tasked by the Air Force to review the program management and the technical reviews and assessments that U.S. Air Force space and nonspace system acquisition programs are required to undergo; assess each review in terms of resources required and its role and contribution, identifying cases where different reviews have common or overlapping goals, content, or requirements; identifying and evaluating options for streamlining, tailoring, integrating, or consolidating reviews of programs to increase cost-effectiveness and to lessen the workforce impact of the reviews as a whole; and recommending changes that the Air Force and the Department of Defense should make. The committee's tasking is shown in Box 1-1.

BACKGROUND

DOD spends over $300 billion per year to develop, produce, field, and sustain weapons systems.[1] Too often, DOD weapons systems programs experience large cost overruns and schedule delays, contributing to a growing loss of confidence

---

[1] See DOD (U.S. Department of Defense), *National Defense Budget Estimates for FY 2009, Updated September 2008*. This amount is the sum of the amounts shown for "Operation & Maintenance," "Procurement," and "RDT&E." Available online at http://www.defenselink.mil/comptroller/defbudget/fy2009/FY09Greenbook/greenbook_2009_updated.pdf. Last accessed May 19, 2009.

BOX 1-1
Statement of Task

The National Research Council (NRC) will

1. Review the prescribed program reviews and assessments that U.S. Air Force space and non-space system acquisition programs in all Department of Defense (DOD) acquisition categories (ACATs) are required to undergo, consistent with the various phases of the acquisition lifecycle, that verify appropriate planning has occurred prior to concept decision, Milestone/Key Decision Point (KDP) A, Milestone/KDP B, and Milestone/KDP C.

2. Assess each review and the resources required to accomplish it, including funding, manpower (people and knowhow), work effort, and time.

3. Assess the role and contribution that each review and the combined reviews make to successful acquisition.

4. Identify cases where different reviews have shared, common, or overlapping goals, objectives, content, or requirements.

5. Identify and evaluate options for streamlining, tailoring, integrating, or consolidating reviews of programs to increase the cost-effectiveness and to lessen workforce impact of the reviews as a whole, including examination and discussion of review processes used by other agencies (such as the National Aeronautics and Space Administration and the Department of Energy), the other military departments (the U.S. Army and the U.S. Navy), and industry.

6. Recommend changes that the Air Force and DOD should make to the reviews of Air Force programs, including review goals, objectives, content, and requirements.

in the DOD acquisition system.[2,3] As reflected in the statement of task in Box 1-1, this study addresses improvements to one of the essential elements of program success—program reviews.

The DOD acquisition decision process is based on phased milestone decisions that are supported by a series of technical and programmatic reviews. These reviews are designed to help program managers (PMs) effectively and efficiently manage the programs and to give executive leadership the information it needs to inform decisions. The formal acquisition decision process in place at the time of the study and used by the committee as the basis for its review is depicted in Figure 1-1.

As noted in the Summary, the May 2003 version of DODI 5000.2 was replaced in December 2008 by DODI 5000.02, shown in Figure 1-2.

The main differences are these: The materiel development decision (MDD) replaces the concept decision (CD); the materiel solution analysis (MSA) phase replaces the concept refinement (CR) phase; the engineering and manufacturing development (EMD) phase replaces the system development and demonstration (SDD) phase, and its two main efforts have been renamed (system integration and system demonstration became integrated system design and system capability and manufacturing process demonstration). Post-CDR assessment replaces the design readiness review.

This formal DOD review process has evolved over the past 60 years, with many of the changes intended to address acquisition program cost overruns, schedule delays, and performance shortfalls in the delivered product, service, or system. Since implementation of the Goldwater-Nichols Act in the late 1980s,[4] the main defense acquisition organizations (e.g., the program management offices) have operated under a tiered decision structure. For large acquisitions, the current policy described in DOD Instruction 5000.1 states that the Under Secretary of Defense for Acquisition, Technology, and Logistics (USD (AT&L)) is the Milestone Decision Authority (MDA) responsible for the overall program:

> The Milestone Decision Authority (MDA) is the designated individual with overall responsibility for a program. The MDA shall have the authority to approve entry of an acquisition program into the next phase of the acquisition process and shall be accountable for cost, schedule, and performance reporting to higher authority, including Congressional reporting.[5]

---

[2]Elizabeth Newell, "GAO: Weapons systems over budget, overdue, underperforming" (April 1, 2008). Available online at http://www.govexec.com/dailyfed/0408/040108e1.htm. Last accessed May 19, 2009.

[3]GAO (Government Accountability Office), *Defense Acquisition: Assessment of Selected Weapons Programs*, GAO-08-467SP, Washington, D.C.: GAO (2008).

[4]U.S. Congress, Goldwater-Nichols Department of Defense Reorganization Act of 1986, Public Law 99-433.

[5]USD (AT&L), The Defense Acquisition System, Department of Defense Directive 5000.1, Washington, D.C.

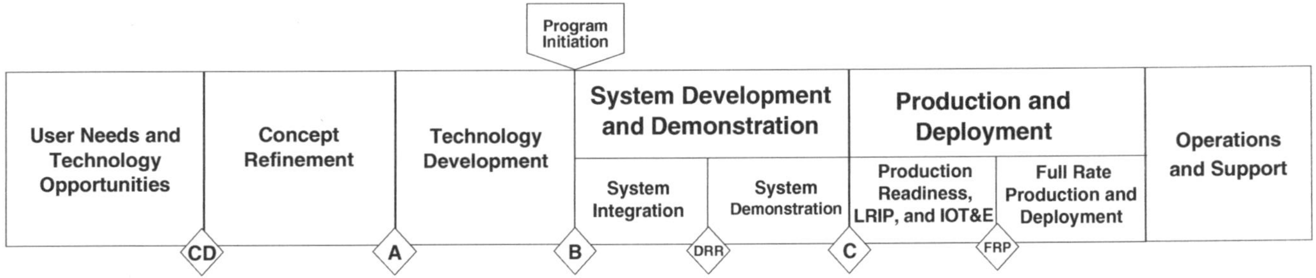

FIGURE 1-1 Major defense acquisition decision reviews and phases. SOURCE: Adapted from John T. Dillard, *Centralized Control of Defense Acquisition Programs: A Comparative Review of the Framework from 1987 to 2003*, NPS-AM-03-003, Acquisition Research Sponsored Report Series, September 2003, Monterey, Calif.: Naval Postgraduate School.

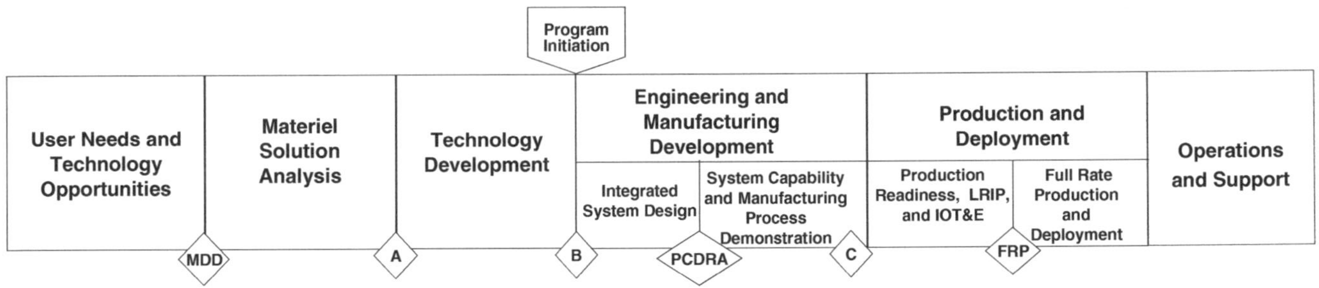

FIGURE 1-2 Revised major defense acquisition decision reviews and phases. SOURCE: Adapted from Figure 1-1 and new DODI 5000.02.

Three levels down the hierarchy, a PM is described as follows:

> The designated individual with responsibility for and authority to accomplish program objectives for development, production, and sustainment to meet the user's operational needs. The PM shall be accountable for credible cost, schedule, and performance reporting to the MDA.

Thus, the PM and MDA share responsibility for development and oversight of a program. Further guidance under DOD Instruction 5000.1 provides as follows:

> There is no one best way to structure an acquisition program to accomplish the objective of the Defense Acquisition System. MDAs and PMs shall tailor program strategies and oversight, including documentation of program information, acquisition phases, the timing and scope of decision reviews and decision levels to fit the particular conditions of that program, consistent with applicable laws and regulations and the time-sensitivity of the capability need.

While the wording above might indicate that the MDA and PM plan jointly or collaborate on program strategy, there are, in fact, both a Service (or Component) Acquisition Executive (SAE) and a Program Executive Officer (PEO) in the hierarchy between them, and direct communication between an MDA and a PM is typically infrequent. The four tiers of major program reporting are shown in Figure 1-3.

Additionally, the Air Force has recently embedded the PEOs and the PMs in a wing/group/squadron framework aimed at aligning acquisition and operational structures.

Figure 1-4 depicts the DOD and Air Force milestone and program review processes. Although changes to both policy and implementation have occurred periodically, the process has its roots in dealing with single programs and/or single systems (platform, weapon, sensor) typically acquired by a single military service. Over the past decade, the emergence of network-enabled programs that require significant interoperability across multiple platforms, weapons, sensor systems, and military services has substantially contributed to complexity and cost of many acquisition programs, complicating program management and the oversight processes.

Beyond decision reviews for major defense acquisition programs at each milestone (A, B, and C), regulations prescribe additional reviews at the Office of the Secretary of Defense (OSD) level for concept (materiel development) decision, design readiness (Post Critical Design Review Assessment (PCDRA)), and full rate production. Before each of these, an overarching integrated product team (OIPT) review is conducted in preparation for the Defense Acquisition Board (DAB) meeting. In preparation for these, a service/component-level review—such

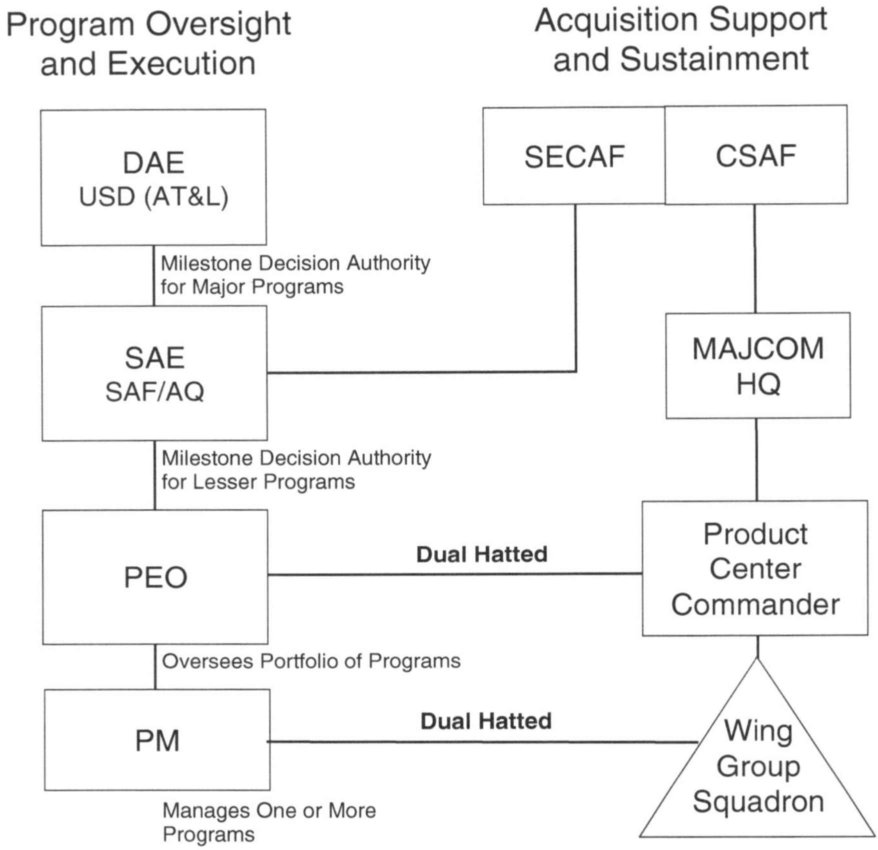

FIGURE 1-3 Four tiers of major program reporting. SOURCE: Committee-generated. DAE, Defense Acquisition Executive; USD (AT&L), Under Secretary of Defense for Acquisition, Technology, and Logistics; SECAF, Secretary of the Air Force; CSAF, Chief of Staff of the Air Force; MAJCOM HQ, Major Command Headquarters; SAE, Service Acquisition Executive, SAF/AQ, Assistant Secretary of the Air Force for Acquisition; PEO, Program Executive Officer; PM, Program Manager.

as an Air Force Review Board (AFRB) or acquisition strategy panel (ASP)—is typically conducted as well. In addition, OSD has implemented program support reviews (PSRs), similar to independent program assessments (IPAs) for space systems, and has directed annual configuration steering boards (CSBs) for programs in the SDD phase. The CSBs are to be chaired by the CAE. At the PEO level, sufficiency reviews are being conducted annually for ACAT I-III programs.

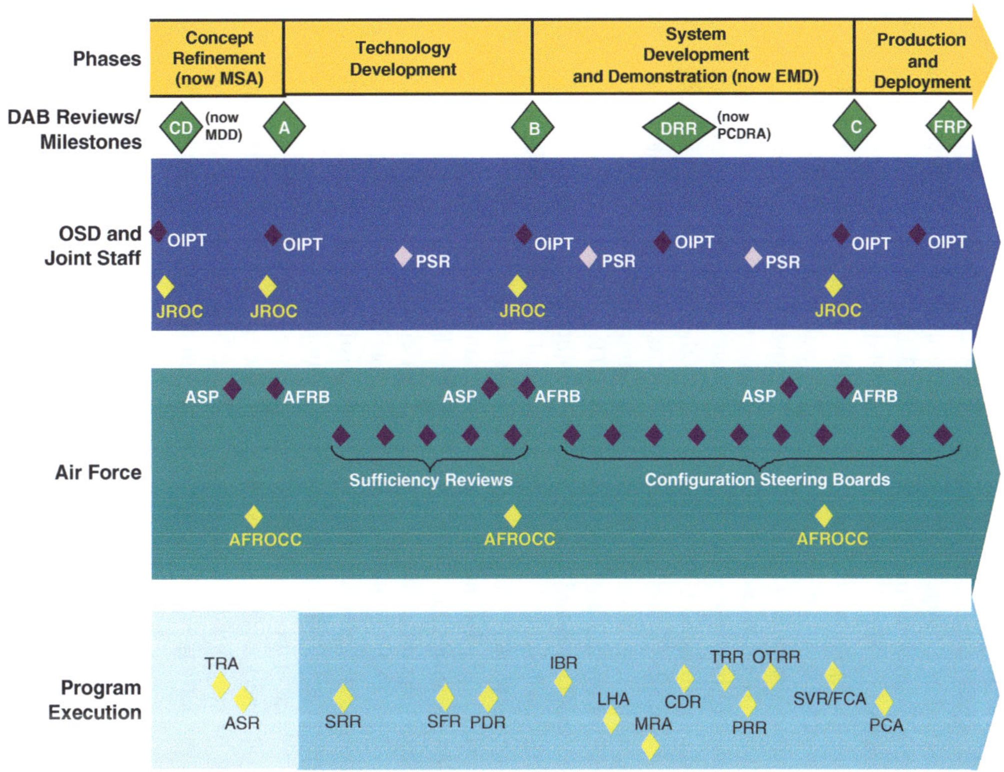

FIGURE 1-4 DOD-Air Force milestone and program review process. For acronyms, see list following the Table of Contents. SOURCE: Adapted from Janet Hassan, Acquisition Chief, Process Office, "Oversight, command and control (OC2)," presentation to the committee on May 7, 2008.

More recent innovations in oversight reviews are specialty reviews, which are assessments (conducted at varying levels) of various aspects of a program, such as logistics health, manufacturing readiness, and technical readiness (maturity).

In the various iterations of the DOD 5000 series regulations governing acquisition programs, both the number and level of reviews have increased substantially, particularly when taking into account the array of prebriefs and informational meetings held in support of the formal reviews.[6] Reviews at multiple levels of the acquisition management hierarchy have increased with each revision of the DOD 5000 series instructions in 2000, 2003, and 2008. The DOD 5000.1 and 5000.2 series of 2000 prescribed six OSD-level decision reviews in the Acquisition Framework for major programs from only four previously (the 1996-era instructions). Its new evolutionary acquisition policy also called for partitioning programs into increments, each requiring its own Milestone B and C reviews. The result was 10 or so reviews in the course of a notional, fully scoped program. More nondiscretionary reviews have since been added in later regulations, and in memoranda from the USD (AT&L) such as the one signed on July 30, 2007, dictating that CSBs chaired by SAEs be conducted annually for major acquisition programs. Similarly, periodic OSD-level program support reviews (PSRs) and assessments of operational test readiness (AOTRs) have arisen to add oversight across functional areas and "improve the probability of program success."[7] Discretionary program-level reviews, such as technical reviews prescribed for systems engineering, were made mandatory in the latest DOD 5000.02 instruction (2008). The net result is a substantial increase in the number and frequency of management reviews at program, service, and OSD levels.

Numerous recent studies[8,9,10] have addressed the cost overruns and delays experienced by DOD acquisition programs over the past few decades. In brief, despite continued attempts to improve the acquisition process, in part through the addition of reviews, acquisition programs continue to experience cost overruns, schedule delays, and/or as-delivered performance shortfalls.

From the perspective of the PM, all of the reviews, both formal and informal, must be supported by the program office, and in many cases the industry partners also participate. Although each individual review is intended to serve a

---

[6]J.T. Dillard, "Toward centralized control of defense acquisition programs," *Acquisition Review Journal*, Defense Acquisition University (DAU), August-December 2005.

[7]Available at http://www.acq.osd.mil/at/initiatives/factsheets/program_support_reviews/index.html.

[8]Assessment Panel of the Defense Acquisition Performance Assessment Project, *Defense Acquisition Performance Assessment Report* (January 2006).

[9]Obaid Younossi, Mark V. Arena, Robert S. Leonard, Charles Robert Roll, Jr., Arvind Jain, and Jerry M. Sollinger, *Is Weapon System Cost Growth Increasing? A Quantitative Assessment of Completed and Ongoing Programs*, Santa Monica, Calif.: RAND Corporation (2007).

[10]GAO, *Defense Acquisition: Assessment of Selected Weapons Program*, Report GAO-08-467SP, Washington, D.C.: GAO (2008).

specific purpose, the overall magnitude of the review efforts not only significantly increases the workload of the program office in terms of direct support, but also diverts attention from day-to-day management of the program.

The committee, in reviewing studies conducted over the past decade, could find no evidence of earlier work that focused on the impact of the overall formal and informal review process on the acquisition system in terms of resources spent by the program office or the effect of diverting a PM's attention from the day-to-day management of his or her programs. Additionally, the unique role that PMs play in the acquisition process requires them to participate in all reviews (and prereviews) with multiple program stakeholders. In brief, only the PM sees and feels the breadth and depth of the review process. For this reason, the committee decided to approach the study from the perspective of the PM, who is a key element in successful program execution.

The committee recognizes the challenges inherent in achieving successful DOD acquisition programs in an increasingly complex and dynamic arena that spans multiple organizations (including industry) and functions that do not easily align. That said, the committee recognizes the opportunity to contribute in a substantive way by examining the expenses a PM incurs from the growing array of program and technical review in terms of time spent supporting reviews and in time lost focusing on program execution.

A key question then is this: Can changes in the number, content, or sequence of program reviews help the program manager execute the program more successfully?

METHODOLOGY

To fulfill the assignments set out in the statement of task, the committee employed a blended research methodology, using four complementary approaches.

Presentations

Data were gathered in the course of four separate multiday conferences. The committee received presentations by PMs and PEOs from the three military departments; industry; DOD overseers, practitioners, process owners, and policy writers; as well as GAO researchers and others who had studied DOD acquisition in a larger context. In addition, some committee members interviewed contributors who were unable to meet with the full committee. The presenters are listed in Appendix B.

Literature Review

In parallel, a committee subgroup accumulated and examined an extensive body of pertinent studies and acquisition reform initiatives within the Air Force, DOD, and other agencies over the last 20 years. The previous studies are listed in Appendix C.

Survey and Other Data

During the compilation and analysis of data from presentations, interviews, and previous studies, it became apparent that there were few data on external program reviews to support this study, particularly items 2, 3, and 4 in the Statement of Task. Consequently, a survey tailored to support this study was developed and beta-tested. The survey was designed to collect information from Air Force PMs and PEOs on external reviews they had experienced. Survey information included quantitative and qualitative data on the impact of external reviews on program execution, including the time and effort spent preparing for, participating in, and following up on actions resulting from such reviews. The survey also asked about PM and PEO assessments of the value of the reviews to them in managing their programs.

The intent of having the survey was to expand the numbers of persons contacted beyond a limited number of interviews and to generate some quantitative data. The committee used the survey data as another form of information to augment its research, interviews, and personal experience. No finding, conclusion, or recommendation in the report is based solely on survey data; rather, the findings reflect what the committee heard from all sources.

Pertinent survey results are discussed in Chapter 2, Findings and Conclusions. The survey can be found in Appendix D, along with a detailed description of how it was developed and conducted and its results. Other data were collected (on a case-by-case basis) from individual programs to fill in missing information about the number and levels of the reviews that are being conducted as part of the current acquisition process described earlier in this chapter.

Comparative Matrix

Lastly, a comparative matrix was constructed as a tool to help identify the number and types of known reviews, their purpose, and their target audiences to identify opportunities for streamlining, integrating, and/or consolidating reviews. The number and types of programmatic and technical reviews are summarized in Chapter 2; a brief description of each review is contained in Appendix E.

Integration and Synthesis of Data

As stated earlier, the committee recognized the substantial body of historical literature and thought addressing the challenges of DOD systems acquisition. Following its review of earlier studies and the series of presentations, the committee spent a significant amount of time discussing how best to respond to the Statement of Task and how best to develop actionable recommendations clearly traceable to study findings. Three early observations substantially influenced those discussions and led to the organization of this report:

1. None of the impressive array of past studies reviewed by the committee approached the acquisition challenge from the perspective of the PM, who is a critical element in the success or failure of a program.
2. The committee's literature review and early interviews indicated that little information existed that would allow it to quantify the resources necessary to accomplish any particular program review as required by item 2 in the Statement of Task.
3. The early round of interviews as well as the collective experience of the committee members led to their sense that there is substantial variance in the conduct and impact of any given review carried out over any set of programs—that reviews are easily influenced by the "personality" or "interest" of the reviewing authority. This observation cast doubt on how well Statement of Task items 3, 4, and 5 could be addressed.

As a result, the committee decided to create and implement a survey to obtain additional information—both qualitative and quantitative, if possible. Additionally, the committee decided to construct a program review matrix (Table 2-1) to present a holistic view of the array of typical reviews (and accompanying prereviews) faced by a program manager. Finally, the committee decided to focus on developing a comprehensive set of recommendations responsive to the statement of task. The intent of the committee is to reflect the perspective of the PM within the larger context of the acquisition environment. The full committee deliberated on the results from all of its information sources to arrive at consensus findings, conclusions, and recommendations.

2

Findings and Conclusions

As summarized in Chapter 1, the committee deliberated on the results from the four data sources to arrive at the findings and conclusions contained in this chapter. The findings and conclusions are based on information derived from the survey responses, interviews with Air Force and DOD personnel and other stakeholders, earlier studies, and the committee members' expertise. The committee spent considerable time deliberating to ensure that the findings indeed represent what it had heard. The conclusions then represent a committee consensus of a generalization of the findings. The recommendations in turn (see Chapter 3) resulted from integrating and synthesizing the findings and conclusions in a way that the committee believed would be most responsive to the Statement of Task.

The matrix of reviews that was constructed to help classify the number and types of known reviews, their purpose, and their target audiences in order to identify opportunities for streamlining, integrating, and/or consolidating reviews is shown in Table 2-1, with further information included in Appendix E. The reviews are of four types (see third column of the matrix):

- *Milestone/programmatic reviews.* These reviews are conducted by the milestone decision authority (MDA) (the Defense Acquisition Executive (DAE) for acquisition category (ACAT) ID programs), with support from the Office of the Secretary of Defense (OSD) staff. They provide the basis for the MDA to decide whether to allow a new program to be initiated or an ongoing program to proceed to the next phase. Of the four types, reviews of this type require by far the greatest number of prereviews (reviews conducted by functional offices and intermediate management levels on the way to the MDA review). These reviews are listed in approximately the order they occur leading up to the ultimate Defense Acquisition Board (DAB) review.

- *Periodic oversight reviews.* These reviews are conducted at a regular frequency—annually or quarterly—to allow the DAE, the Service Acquisition Executive (SAE), or the program executive officer (PEO) to monitor progress. They are listed alphabetically.
- *Ad hoc reviews.* These reviews are initiated by an organization outside the program office either in response to a problem or, in the case of some major programs, periodically (e.g., by the GAO). They are listed alphabetically.
- *Technical or engineering reviews.* In contrast to the external reviews mentioned above, these reviews are generally conducted internally by the program itself, with some support from Service functional assets. They are a principal means by which the program office manages the technical and fiscal execution of the program. They are listed in approximately the order they occur over the life of the program.

The matrix (Table 2-1) summarizes the various types of reviews typically conducted over the lifetime of a program. It names 31 formal reviews of four types, although not all of them or even all types may be conducted for every program. At least four of the reviews require unique documents and 10 have been identified by the committee as duplicating or partially duplicating other reviews.

The matrix does not list every possible ad hoc review. Nor does it list the prereviews or prebriefs generated by these formal reviews, since it was not possible to determine the number of such reviews or prebriefs. Because no data or metrics are required or collected on reviews by DOD in general, or by the Air Force in particular, it was not possible to determine the overlap or duplication of different reviews. (Of note, although no historic information on the number of prereviews or prebriefs is available, interview comments and the committee members' experience lead one to believe that major reviews at the level of the Joint Requirements Oversight Council (JROC), DAB, and PEO are generally accompanied by multiple (3 to 5) prebriefs or prereviews.)

In the program review process depicted in Figure 1-4, it can be seen that there are 6 MS or MDA reviews in a typical program, plus 6 overarching integrated product teams (OIPTs) reviews before those, plus 6 Air Force Review Board (AFRB)/acquisition strategy panel (ASP) reviews before each of those, plus 10 or 11 mandatory technical reviews at the program level, plus 4 periodic program support reviews (PSRs) for each phase of the life cycle, plus one operational test readiness assessment (OTRA) before IOT&E, plus annual configuration steering board (CSB) reviews (notionally 7), plus annual PEO-level sufficiency reviews (notionally 7), plus at least 2 technology readiness assessments/manufacturing readiness assessments (TRAs/MRAs). This suggests in excess of 50 reviews just at the program level and above without counting

TABLE 2-1 Program Review Matrix

Review	Name	Type	Purpose	Frequency	OPR/ Customer
ASP	Acquisition Strategy Panel	Programmatic	Formalize acquisition strategy	MS	SAF/AQ
AFROCC	Air Force Requirement for Operational Capabilities Council	Programmatic	Validate requirements	MS	A5R/CSAF, SECAF
JROC	Joint Requirements Oversight Council	Programmatic	Prioritize Joint requirements	MS	VCJCS/DAB
AFRB	Air Force Review Board	Programmatic	Ready for MS or OSD	MS	SAE
IPA	Independent Program Assessment	Programmatic	Risk assessment	MS	SMC/MDA
PSR	Program Support Review	Programmatic	Risk assessment	MS	Director SSE/ USD(AT&L)
IIPT	Integrating IPT	Programmatic	Support OIPT	MS	PM/OIPT
OIPT	Overarching IPT	Programmatic	Ready for DAB	MS	Dir PSA, DASD(NII)/ DAB
DAB	Defense Acquisition Board	Programmatic	Advise USD(AT&L)	MS	USD(AT&L)
CSB	Configuration Steering Board	Periodic oversight	Review requirements changes	Annual	SAE/ USD(AT&L)
DAES	Defense Acquisition Executive Summary	Periodic oversight	Ongoing performance	Quarterly	PM/ DUSD(A&T)
PEO/SR	PEO Sufficiency Review	Periodic oversight	Executability confidence	Annual	PEO

Presenter	Stakeholders	Prereview Briefings	Product/Output	Unique Documentation?	Duplication
PM	SAF/AQ	NDA	Approve acquisition strategy plan	Yes	Partially
A5 staff, MAJCOM representative	Air staff, AF Secretariat	FCBs	Validate requirements	No	NDA
J8 director, Service representative	Services, warfighters	Joint Requirements Board, Panel	JROCM	Yes	No
Requirements representative	PM, user, OSD, service, KTR	Unknown	Readiness to proceed	No	No
PM, staff	DSAB, staff support	Unknown	Briefing and report	No	Partially
PM, PO & KTR SMEs	Dir SSE, USD(AT&L), PM	Coordination meetings	Brief findings, recommendations	No	Partially
Varies	PM, user, OSD, service, KTR	None	Report to OIPT	NDA	Partially
PM; OSD staff	PM, PEO, OSD, Joint Staff	IIPT or equivalent, staff	Report w/ recommendations	No	No
OIPT chair, PM, others	USD (AT&L), DAB members	OIPT	Acquisition Decision Memorandum	Yes	No
PM	SAE, USD(AT&L), JCS	Unknown	Approve requirements change	Yes	NDA
PM (with SAE)	PM, SAE, OSD staff	Unknown	Action items	Yes	No
PM, SMEs	PEO, PM, ACE, SMEs	Unknown	Briefing, PoPs update	No	NDA

continued

TABLE 2-1 Program Review Matrix (*continued*)

Review	Name	Type	Purpose	Frequency	OPR/ Customer
AFAA	AF Audit Agency	Ad hoc oversight	Varies	Varies	AFAA/SECAF
DOD IG	DOD Inspector General	Ad hoc oversight	Varies	Varies	DoD IG/ SecDef
GAO	Government Accountability Office	Ad hoc oversight	Cost, schedule, performance	Varies	GAO/Congress
Other ad hoc		Varies	Varies	Varies	Varies
TRA	Technology Readiness Assessment	Technical	Executability confidence	MS	DDR&E/MDA
ASR	Alternative System Review	Technical	Ready for technology development	MS A	Engineering/ PM
SEAM	Systems Engineering Assessment Model	Technical	Validates SE process	Varies	AFCSE/PM
SRR	System Requirements Review	Technical	Executability confidence	One time	Engineering/ PM
SDR	System Design Review	Technical	Replaced by the SFR	One time	Engineering/ PM
SFR	System Functional Review	Technical	Ready for prelim design	One time	Engineering/ PM
PDR	Preliminary Design Review	Technical	Ready for detailed design	One time	Engineering/ PM
IBR	Integrated Baseline Review	Technical	Align program expectations	Varies	PM
LHA	Logistics Health Assessment	Technical	Logistics health	Unknown	Unknown; currently AAC
MRA	Manufacturing Readiness Assessment	Technical	Executability confidence	One time	DDR&E/MDA
CDR	Critical Design Review	Technical	Ready for fabrication	One time	Engineering/ PM

Presenter	Stakeholders	Prereview Briefings	Product/Output	Unique Documentation?	Duplication
PM, staff	SECAF, staff, AFAA office	Coordination meetings	Report	No	Partially
PM, staff	Varies - wide-ranging	Coordination meetings	Written report w/ recommendations	No	Partially
Varies - often PM or rep	Varies - wide-ranging	Coordination meetings	Written report w/ recommendations	No	Partially
Varies	Varies	Unknown	Varies	Varies	Yes
AF ST&E	PM, S&T, DDR&E, DAE	Unknown	Tech Readiness Levels/Plan	NDA	Yes
PM, PO & KTR SMEs	PM, User, KTR	Unknown	Rationale for preferred alt	No	No
PM/ Engineering	AFCSE, PM	None	Assessment reports	No	Partially
PM/ Engineering	Engineering, user, KTR	Unknown	Satisfy exit criteria	No	No
PM/ Engineering	PM, contractor	Unknown	Satisfy exit criteria	No	No
PM/ Engineering	Engineering, user, KTR	Unknown	Satisfy exit criteria	No	No
PM/ Engineering	Engineering, user, KTR, DAE	None	Satisfy exit criteria	No	No
PM	PM, contractor	NDA	Mutual understanding of program baseline	No	Partially
Unknown	PM, user, AF logisticians	Unknown	Log/sustainment assessment	Unknown	Partially
PM/ Engineering	PM, S&T,	Unknown	Mfg Readiness Levels/Plan	NDA	NDA
PM/ Engineering	Engineering, User, KTR, DAE	None	Satisfy exit criteria	No	No

continued

TABLE 2-1 Program Review Matrix (*continued*)

Review	Name	Type	Purpose	Frequency	OPR/ Customer
TRR	Test Readiness Review	Technical	Ready for testing	One time	Engineering/ PM
PRR	Production Readiness Review	Technical	Ready for production	MS C, FRP	Engineering/ PM
OTRR	Operational Test Readiness Review	Technical	Approve program OT readiness	One time	SAF/AQ
SVR	System Verification Review	Technical	Verifies performance	MS C	Engineering/ PM
FCA	Functional Configuration Audit	Technical	Satisfies contract	One time	Engineering/ PM
PCA	Physical Configuration Audit	Technical	Satisfies contract	One time	Engineering/ PM

Frequency: MS, this review takes place in support of or in conjunction with each formal program milestone (A, B, C, etc.); MS A, this review takes place in support of or in conjunction with formal program milestone A; MS C, this review takes place in support of or in conjunction with milestone C; MS C, FRP, this review takes place in support of or in conjunction with milestone C and the full rate production decision.
Duplication: A review/meeting exactly the same as another review/meeting. By this definition, the use of the same material at more than one meeting does not constitute a duplication of the meeting. The

the prebriefs to staff functional principals and others as noted in the preceding paragraph.

Although the data and information received did not allow the committee to determine specific resources required to accomplish each review, answers to Question 1.5 in the survey and information gathered from the interviews indicated that 10-30 percent of a PM's time is spent supporting reviews. The committee found that there is little consistency in the way reviews are conducted and concluded that opinions on the contribution of specific reviews to successful acquisition varied widely.

The committee believes that the Air Force could improve the effectiveness of its program review effort and reduce the burden on PMs by thoughtfully combining and scheduling reviews. The committee looked at the policies and processes of the National Aeronautics and Space Administration (NASA), the Department of Energy (DOE), industry, and the other military services. Although review

Presenter	Stakeholders	Prereview Briefings	Product/Output	Unique Documentation?	Duplication
PM/ Engineering/ Test	Test IPT, User, KTR, DOT&E	None	Prepared for testing	No	No
PM/ Engineering	KTR, Subs	None	Requirement met, production ready	No	No
PM	SAF staff, DOT&E	NDA	ASAF(A) approval of program's OT readiness	Yes	Partially
PM/ Engineering	User, KTR	None	Requirements met, funding adequate	No	No
PM/ Engineering	PM, Engineering, KTR	None	Audit report	NDA	No
PM/ Engineering	PM, Engineering, KTR	None	Audit report or completion	NDA	No

degree of duplication shown in the Duplication column is subjective and is based on review documentation, interviewee comments, and survey results. Given that no review, let alone its process owner, requires any data collection on duplication, coordination, value added, and so on, the conclusions are not rigorous. Review duplication attributes: Yes, the review was judged to be a duplication; No, the review was judged not to be a duplication; Partially, the meeting was judged to be a partial duplication; and NDA, the review is too new or had no data available to allow a judgment on duplication.

structures and practices exist, the differences in implementation make cross-community consolidation or streamlining a significant challenge. The information collected from experienced PMs in the U.S. Army and the U.S. Navy, as well as the U.S. Air Force, and from industry contributed to the set of best practices reported in the recommendations.

REVIEWS HAVE BOTH BENEFITS AND COSTS

The committee found that program and technical reviews have both benefits and costs. In general, reviews provide technical and programmatic support to successfully execute acquisition programs, to inform decisions, to share awareness, and to engender program advocacy. In their answers to Survey Question 2.3, PMs said that reviews facilitated program execution as well as problem discovery and resolution at all levels of the acquisition enterprise, including industry. That said, support for the increasing number of program reviews and associated prebriefs is costly, as is the lost productivity and attention of both government and industry. For example, DODI 5000.02 of December 2008 calls for systems engineering across all phases and mandates technical reviews. The new regulation mandates competitive prototyping of the system or key system elements during the technology development (TD) phase, and a preliminary design review (PDR) must be conducted for each of the candidate designs, followed by a PDR report to be provided to the MDA either in the TD phase supporting milestone B or, if afterward, a separate MDA review during the engineering and manufacturing development (EMD) phase.

Although some such overhead is undoubtedly necessary, excesses can distract PMs from their primary focus, which is managing their program's technical and business progress. The findings and conclusions of the committee focus on two key areas: The first is the execution of reviews (sequencing, timing, participation), and the second is their planning.

EXECUTION OF REVIEWS (SEQUENCING, TIMING, PARTICIPATION)

The committee's investigation confirmed that the number of reviews is growing. The committee recognizes that such reviews are generally valuable for sharing knowledge and serve as "stage gates" for the governance and control of programs, including managing risk. While doing this, however, they most certainly add to program costs, and they also draw management's attention from the main program effort. Of most concern is that the proliferation of reviews does not appear to have had a positive effect on program cost and schedule outcomes. The committee also realized that there is a significant amount of preparation and coordination for reviews, both vertically in their conduct at multiple levels of responsibility and horizontally across adjacent staff offices. As such, merely depicting the individual reviews does not sufficiently capture the amount of time and effort spent by staff in preparing for and coordinating the reviews.

The committee found that the many disparate concerns of higher-level staffs had an impact on the program manager. For ACAT I programs, many of the written-out responses[1] to Survey Questions 4.1-4.4 described DOD staff as a stove-piped bureaucracy, where domain "czars" have purview over a breadth of programs (by virtue of the OIPT structure or their membership on the DAB) but are not horizontally integrated from the standpoint of knowledge sharing or synergy. This means not only that the PMs have to prepare separate information

[1]The qualitative write-in comments from the PMs have not been included in this report because several of the comments were written in such specific detail that the authors of the comments could be identified by people familiar with their programs. This would violate the privacy protection methodology for the survey, so the data are not included.

briefs for these higher-ups, but also that the information provided to them is not fully integrated across these domains by the OSD staff for optimal decision making by the MDA.

Programs and technical reviews are often not being optimally synchronized with program events. Speakers gave examples of requirements reviews being conducted after contracts had been awarded and other similar occurrences of inappropriate sequencing. The committee was briefed on emerging acquisition policy with its emphasis on pre-Milestone A activities to ensure a better understanding of alternative technical requirements. The committee believes there is ample opportunity for the Air Force and DOD to improve the timing of these reviews and even to consolidate reviews at the service staff level and below to eliminate redundancy and mitigate the associated burdens and costs.

Finding 1a. A significant share of a PM's time is spent on preparing for and participating in milestone and other reviews.

Finding 1b. The large number of reviews diverts a PM's attention from the execution of his or her program.

Finding 1c. Reviews impose significant costs on program leadership teams.

The committee interviewed a number of PMs to obtain their views on the time they spent and the costs associated with carrying out these various reviews. It also sought their assessment of how the conduct of these reviews affects their ability to carry out the day-to-day tasks associated with their programs. The PM survey also asked respondents to address these issues. In every case, those interviewed or surveyed cited significant costs in terms of money and time to carry out the reviews, and most of them also noted an adverse impact on their ability to carry out other PM responsibilities. Sixty-nine percent of survey respondents said they were working 51 or more hours per week (Survey Question 1.4) and on average were able to spend only 46 percent of that time managing their program at wing level or below (Survey Question 1.5). About 20 percent of their time was spent reporting up the chain of command above the wing level (Survey Question 1.5). In addition, many of the respondents cited the need to dedicate hundreds or thousands of program staff and contractor hours to carry out the reviews, with concomitant money and time lost from their everyday duties (based on write-in responses to Survey Questions 3.11-3.14 and 3.26-3.29).

Finding 1d. Many higher-level reviews do not contribute to program execution in proportion to their expenditure of time and effort.

While those interviewed identified benefits they received from reviews that were effectively carried out, a number of survey respondents and interviewees said they received no significant benefit or improvement to program cost, schedule, or performance (CSP) from the effort. For example, 29 PMs identified what they considered the least beneficial (from the standpoint of program CSP) higher-level review they had been involved in. Eighty percent of them cited no positive impact on CSP from this "least beneficial review" (based on written-in responses to Survey Question 3.24). Some even went so far as to say that the review had actually had a negative impact on CSP by delaying program schedule or increasing program costs in the case of reviews that were ineffectively carried out (based on written-in responses to Survey Question 3.25). It should be noted that the PMs were not questioned on ancillary benefits such as sponsorship, which they may not have viewed as a direct program benefit.

Finding 1e. No one in the Air Force or OSD is responsible for monitoring the number of program and technical reviews or the workload they give rise to or their cost, effectiveness, and cumulative impact on PMs.

Based on the committee's review of the OSD and Air Force organizational structures and its discussions with senior DOD leadership, no one is responsible for monitoring the direct costs of program and technical reviews, in terms of the time, personnel usage, extra costs, and effects on contractors and the PM.

Conclusion 1. Many reviews add little value and others do not add value in proportion to the effort required. Reducing the number of such reviews or combining them could increase the time available to the PMs to more effectively manage their programs.

Finding 2. The sequencing, timing, and frequency of reviews are often not tied to the program schedule in a manner that most effectively supports the program and its execution.

Forty-three percent of survey respondents commenting on the least beneficial reviews suggested that those reviews could be more effective if they were conducted less frequently and at a more appropriate time in the program's life cycle (generally earlier) (Survey Question 2.6).

Conclusion 2. Reviews could be more effective if they were sequenced and timed to provide the information needed for program execution. Chapter 3 describes an approach for beginning to achieve this result.

Finding 3a. The program review principals, key stakeholders, and subject matter experts do not always attend program reviews.

Finding 3b. Program and technical reviews are often not attended by the right personnel or, in some cases, are attended by too many personnel.

Many survey respondents noted that more effort should be given to ensuring that the right subject matter experts and appropriate senior officials attend program reviews and that the number of attendees be limited to those who can add value to the meeting (based on responses to Survey Questions 2.6 and 3.15 and written-in responses to Survey Questions 3.30-4.4).

Conclusion 3. Required attendance at program review meetings is not clearly communicated nor is it effectively controlled.

Finding 4a. For some reviews the number of actual reviews and preparatory reviews is excessive and the reviews do not contribute value to the program's management.

Many of the PMs, both in the survey and in interviews, stated that a proliferation of meetings and premeetings was taking time away from the management of their programs (based on responses to Survey Question 2.6 and written-in responses to survey Questions 3.10 and 3.25). Commenting on the reasons for this increase, a few PMs mentioned the Integrating Integrated Product Team (IIPT) and the NII/AT&L structure. The elimination of IIPT reviews was cited as one factor that led to the need for more individual premeetings with the Joint Staff, program management offices, and OSD, where previously only one meeting was needed. Another reason cited was the recent NII/AT&L reconfiguration that resulted in the sharing of responsibilities between these two offices. Survey respondents cited the amount of time now taken up by such premeetings. One respondent, for example, said "The problem isn't the review . . . it's the numerous premeetings needed to get to the review."

Finding 4b. Program managers are spending time on multiple reviews with similar objectives.

Many survey respondents believed that selected reviews could be combined (Survey Question 2.7).

Conclusion 4. Streamlining or combining reviews and their associated prebriefs in both the vertical and horizontal directions could increase efficiency.

PROGRAM REVIEW PLANNING

The committee found that in many cases, despite published guidance advocating proper planning, reviews were incompletely planned or conducted. The potential value of a review cannot be fully realized if objectives are not clearly specified or if the right persons are not in attendance. What information a particular review is supposed to contain or what areas a particular review is supposed to cover are often ill defined or based on presumptions of agendas or issues of the day. Often there are no metrics for assessing the effectiveness of a review, and this can lead to perceptions of disruption and uncertain value. On the other hand, many PMs found that the program support reviews (PSRs) and the independent program assessments (IPAs) were well planned, that they added value, and that they contributed positively to program management.

PSRs and IPAs are comprehensive in nature; have well-defined processes, outcomes, and metrics; are socialized with PMs and staff; are conducted by subject matter experts; and are well documented. A review process that makes timely use of PSRs and IPAs might help to limit the number of reviews needed across the bureaucracy to gather information and make decisions.

Finding 5. The purpose, scope, information needs, key issues, and expected outcomes of many reviews have not been specified.

Survey respondents mentioned a number of ways to improve reviews. Suggestions for improvement to address the problems cited above included narrowing the review's focus or changing its charter (based on responses to Survey Questions 2.6 and 3.15 and on written-in responses to Survey Question 3.30). PMs noted that they would be asked about issues that had not been previously identified for discussion and would then be required to spend countless hours after the review trying to respond to them. They also said there was no standard approach to how the reviews should be conducted.

Conclusion 5. It is important that program review planning is accomplished in a thoughtful, purposeful manner with a standard approach in order to firmly address the need for communication of the expectations and outcomes.

Finding 6. Reviews focus on a single system instead of on a complex system of systems of which the single system is a part. Further, reviews that attempt to address programs from the larger system-of-systems perspective are often unable to cope with the complex interfaces among programs.

Seventy percent of the ACAT I PMs who responded to the survey characterized the amount of external interface of their programs with other efforts as extensive (Survey Question 1.13, ACAT 1 PM responses only). This answer—combined with survey written-in responses and PM discussions with the full committee noting that some reviews did not take into account connections with and dependencies on other programs for mission accomplishment—allows concluding that the current acquisition and program review process has not adapted to the evolution of simple systems into systems of systems and fails to take into consideration the additional complexity and interrelationships necessary for effective program management in this new environment (based on written-in responses to Survey Questions 3.10, 3.25, and 4.1-4.4).

Conclusion 6. Program review format and design need to reflect the greater complexity and interrelationships inherent in many current Air Force programs to ensure that a system of systems works across organizational constructs.

3

Recommendations

As discussed briefly in the section on methodology in Chapter 1, the findings and conclusions set forth in Chapter 2 serve as the basis for the recommendations contained herein that would allow the Air Force and the DOD to achieve meaningful and constructive change. The data collected by the committee—the sum of the literature review, survey results, interview comments, and committee experience—suggest that substantial variances exist in the planning and execution of program reviews as they apply to multiple programs. In brief, each program is different and each review is conducted in a different manner, with different participants and different results. The same type of program review applied to two different programs may have differing results. Additionally, since neither the data nor the time available permitted a detailed, direct, one-on-one response to items 2-4 in the Statement of Task, it is difficult to make recommendations that might apply to specific program reviews.

As a result, the committee decided to focus its response to the Statement of Task in the form of a recommended approach—a set of principles that might form a core set of best practices that applies to each specific program as well as to the coordination and synchronization of all reviews—with the goal of increasing the effectiveness and efficiency of program reviews and decreasing the burden on the program manager. The five recommendations apply to the execution of program reviews as addressed in the Statement of Task. Further, although there are not enough specific data to permit a quantified response to the key question raised in the summary—namely, Can changes in the number, content, sequence, or conduct of program reviews help the program manager more successfully execute the program?—the findings, interviews, and survey results gathered by the committee indicate that addressing the administrative issues surrounding reviews can have a very positive impact on the ability of PMs to execute their programs more successfully.

Following the recommendations, the committee provides comments regarding an approach for their implementation. Table 3-1 lists the conclusions from Chapter 2.

Recommendation 1. To ensure that they possess a common understanding of the intent, scope, and output of reviews, the Air Force acquisition and requirement communities at all levels should engage in timely planning for program reviews that results in clear, comprehensive, measurable objectives.

This recommendation, based principally on Conclusions 1, 5, and 6, reflects the committee's discussion and desire to ensure that each program review is planned and conducted with thoroughness and precision to achieve success. The committee acknowledges the significant challenges posed by Conclusion 6 in terms of the comprehensiveness of planning and processes needed to ensure proper system-of-systems integration. This may well be an area for future study.

To execute this recommendation, a governance process directed by the Service Acquisition Executive (SAE) should be implemented to synchronize and execute reviews at each level of organization. The governance structure must have an owner of the review process that encompasses all reviews captured within that structure, all policies issued in connection with those reviews, and control of the proliferation of reviews, including pre- and postreview mechanisms.

Engagement in program review planning is not consistent throughout the Air Force. Such planning must be deliberate and should be communicated to the PM

TABLE 3-1 List of Conclusions

Conclusion	Description
1	Many reviews add little value and others do not add value in proportion to the effort required. Reducing the number of such reviews or combining them can increase the time available to the PMs to more effectively manage their programs.
2	Reviews could be more effective if they were sequenced and timed to provide the information needed for program execution.
3	Required attendance at program review meetings is not clearly communicated nor is it effectively controlled.
4	Streamlining or combining reviews and their associated prebriefs in both the vertical and horizontal directions could increase efficiency.
5	It is important that program review planning is accomplished in a thoughtful, purposeful manner with a standard approach in order to firmly address the need for communication of the expectations and outcomes.
6	Program review format and design needs to reflect the greater complexity and interrelationships inherent in many current Air Force programs to ensure that a system of systems works across organizational constructs.

and the office of primary responsibility (OPR) for program reviews well in advance of the reviews. Direction for a review should include, at a minimum, the stated objective(s) of the review; the metrics for those objectives; the materials the PM is expected to supply to the review team, including supporting material; and the criteria for success for the program review. To this end, the committee suggests that the SAE review the application of leading industry standards developed by internationally recognized project and program management associations. Such standards have met the test of benchmarking in many industries, are globally applicable, and can be easily adapted.

Following the program review, the chair of the review team should issue a report giving the PM the findings that require the PM to take corrective action and recommendations for further actions. Further, the report should compile some lessons learned from carrying out the review.

To complete the review process, the PM should file a closeout report with the chair of the review detailing implementation of the corrective action plan and recommendations. The closeout report should note open items, closed items, items still in process, and project issues or risks that have been encountered or are predicted when the report is filed. The report should be taken into account in any follow-on program review.

The committee further recommends that the SAE track the various metrics outlined by the review committees to determine if the reviews are having a significant impact on program performance. Such data could be used to improve the program review process as well as focus reviews on areas of concern.

Recommendation 2. The SAE should develop a plan for the timely, synchronized execution of all program reviews. The plan should align with program decision milestones and decision points.

This recommendation is based principally on Conclusions 2, 4, and 6. Its goal is to coordinate and synchronize the array of program reviews both horizontally and vertically across the department.

The number of reviews preceding the decision points and milestones should be minimized and those that are held are overseen to ensure the content is pertinent. This will reduce the burden on the program and assure that the reviews bring value to the program. In some cases, the alignment of those reviews may allow them to be consolidated and moved to a more appropriate time in the life cycle of the program. Program reviews are aligned with the decision points and milestones to ensure better program execution. Properly synchronized reviews should bring fewer schedule delays and reduce costs, since early identification of issues and risks should allow the PM to institute better planning and handling strategies.

The elimination of some reviews or the combination of others will reduce costs as well as reduce the burden on the PM and program management staff, allowing greater focus on the program and its execution. This idea is elaborated in the next section, "Implementing the Recommendations."

Recommendation 3. Before creating or approving a new review, the SAE should compare its objectives with those of existing reviews to determine whether one of the latter could accomplish or incorporate those objectives.

This recommendation reflects discussions based on Conclusions 1, 2, 4, and 5 and the committee's sense that the burden on the PM is exacerbated by additional reviews that arise during the course of program execution and are not necessarily coordinated or synchronized with previous or future reviews.

Evidence from the survey and the interviews indicates that many PMs believe that the number of program reviews is a burden to their program and that adding more would only increase that burden.

Recommendation 1 will keep this from happening as will the assessment process called for in Recommendation 3, part of which should determine if the objective of the proposed review could be achieved within an existing review. The assessment process should also determine whether broadening the stakeholders for a given review would do more to accomplish an objective than instituting additional program reviews. All review process determinations should keep Recommendation 2 in mind to ensure that reviews are timely, synchronized, and add value to the program.

The same criteria established for a review should be applied to any and all prereviews, including those requested before reviews required by the Program Executive Office, the Air Force, or the Office of the Secretary of Defense (OSD). If possible, all stakeholders should work together to consolidate the prereview process. The committee understands that many stakeholders want the program review to be successful and that having some prereviews might be worthwhile. To that end, the committee encourages the Air Force to establish guidance for managing a prereview so that it has a minimal impact on the schedule and cost of the program and the program management staff. The committee also recommends that the Air Force encourage OSD to do the same, particularly regarding prebriefs.

Recommendation 4. The OPR should staff the review team with recognized subject matter experts.

This recommendation reflects committee discussions related to Conclusions 3 and 6. The committee was somewhat surprised to learn that many reviews are conducted without the "right" people present. This raises two issues. First, recognized subject matter experts need to be identified. Second, the experts must participate in the program review for their expertise to be of full value to the program.

To act on this recommendation, the OPR should maintain a roster of experts in standard technical areas, taking into account that back-ups will be needed for any given program review to guarantee that an expert is available to attend that review. Further, a process guidance document should be prepared that would provide direction for the selection of subject matter expert teams, their formation, and their use.

The objective(s) of the program review should be considered when staffing the review team. Moreover, what counts is to ensure continuity of effort—that is, the availability of the subject matter experts not only during the review but also for the periods before and after the review.

Recommendation 5. The OPR conducting the review should ensure that all review outputs are documented, including root causes if any have been identified, and provide recommendations that can be acted upon by the PM, the program management office, or other program stakeholders.

This recommendation is based on discussion related to Conclusion 5 and reflects the committee's view on the importance of proper documentation for follow-through and for sharing lessons learned across programs.

The output of the program reviews is sometimes not fully captured. The committee notes that it is a best practice to capture lessons learned, identifying the root cause of problems and encountered risks in program management, as well as to document findings, observations, and recommendations made during a program review. Best practices give the PM and his or her management staff a roadmap to improvement or recovery. They also provide access to technical experts who can lend their knowledge at critical junctures in program development and execution, carry out monitoring and reporting functions, and serve as a vehicle for transferring and disseminating the body of lessons learned and the knowledge of senior Air Force technical and managerial officials.

Documenting the output of reviews stimulates open communication and builds an atmosphere of trust that will lead to participation in future program reviews. Documented feedback will mean that decisions can be tracked and implemented as well as communicated to the decision makers. Further, the management and execution of programs will continue to improve as program personnel learn from the experience of the review team.

The committee recommends two ways to capture the information conveyed during the review. The first way is for the review team to write a report about the review. The second is to create a database for storing lessons learned and sharing them with the rest of the Air Force and others. The database would allow the lessons learned to benefit an audience beyond the immediate PM, and it would begin the virtuous cycle of review and improvement that the Air Force is seeking. The committee suggests that this database of lessons learned be owned and administered by the SAE. To be useful, it needs to be searchable and updated regularly.

In summary, the committee believes that implementing these five recommendations will improve the efficiency and effectiveness of program reviews. Together, the recommendations form a "gold standard" for conduct of reviews.

IMPLEMENTING THE RECOMMENDATIONS

As illustrated in Chapter 1, a typical ACAT I program review structure might look like Figure 3-1.

Note that Figure 3-1 reflects a milestone-driven process and does not show all the numbers of prereview briefings associated with each review. Recently revised DOD Instruction 5000.02 does not appreciably change the focus on milestones but it introduces more oversight reviews with the aim of achieving better acquisition results.

As stated earlier, the committee found that, although it is critical for the assessment of each individual review, information on costs, manpower, effort, content, objectives, and so on, was not available at every level. Despite the sparseness of information, the committee's interviews, findings, and survey results strongly suggest that better administration of the review process—including synchronizing, combining, and aligning reviews—would alleviate the burden on the PM and help him or her achieve program success. Given the existing DOD organizational hierarchy and culture, the committee believes that the areas with most potential for consolidation and streamlining are the various external reviews and assessments, such as the configuration steering boards (CSBs), PEO sufficiency reviews (PEO/SRs), and prebriefs to the OSD staff, that are carried out for, say, a year before a milestone or other major review. The content of CSBs and PEO/SRs might be combined with phase milestone reviews or intermediate phase reviews. As for prebriefs, the reinstatement of the IIPT and strict enforcement (by the Air Force and OSD) of limits on OSD staff prebriefs to that forum would do a lot to decrease the number of prebriefs before the DAB milestone review and other DAB-level reviews.

Figure 3-2 represents a nominal approach to synchronizing and integrating a series of program reviews aligned to major program milestones. Given the sheer number and frequency of program and technical reviews, combining at least some of them is seen by the committee as a way to improve effectiveness and efficiency while still satisfying the decision support needs of multiple stakeholders.

The committee believes that combining and synchronizing reviews in this way should significantly improve program management (and governance) efficiency and effectiveness.

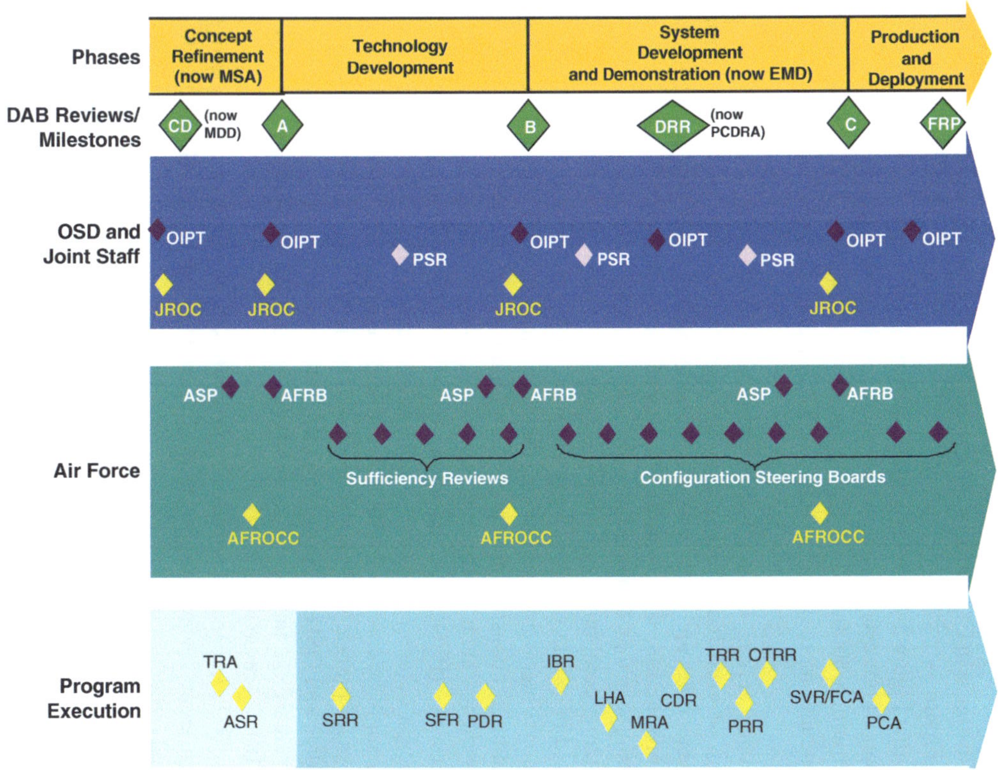

FIGURE 3-1 DOD-Air Force milestone and program review process. For acronyms, see list following the Table of Contents. SOURCE: Adapted from Janet Hassan, Chief, Acquisition Process Office, "Oversight, command and control (OC2)," presentation to the committee on May 7, 2008.

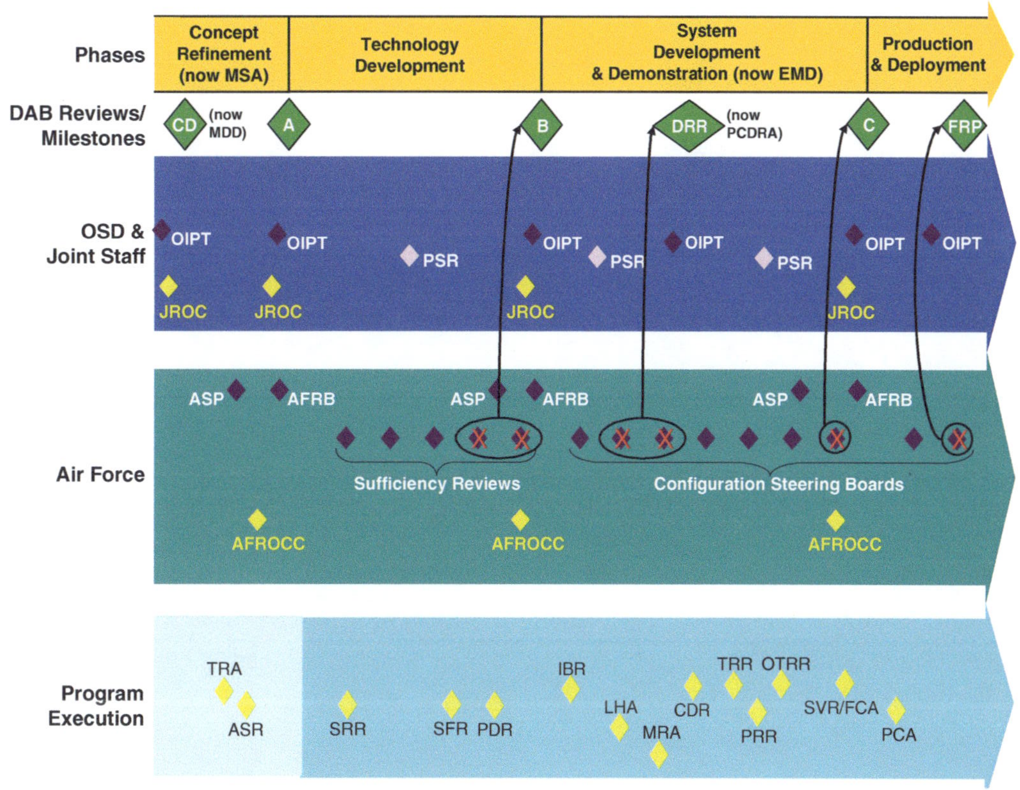

FIGURE 3-2 Areas for potential consolidation and streamlining. SOURCE: Committee-generated; modification of Figure 3-1.

Appendixes

Appendix A

Biographical Sketches of Committee Members

Rand H. Fisher, *Chair,* is currently senior vice president for systems planning and engineering for The Aerospace Corporation. Before he retired from the U.S. Navy, RADM Fisher served concurrently as director of the Communications Acquisition and Operations Directorate within the National Reconnaissance Office, commander of the SPAWAR Space Field Activity for the Space and Naval Warfare Systems Command, naval program executive officer (PEO) for Space Systems, and director of the Transformational Communications Office. He previously served as commander of the Naval Air Warfare Center Weapons Division in China Lake, California, and assistant commander for test and evaluation, Naval Air Systems Command. RADM Fisher has also served as the Space and Naval Warfare Systems Command's research and development program manager; as director of the Systems Program Management Division; as lead systems engineer for the Naval Space Technology Program; as deputy program manager and then as major program manager, Special Systems Program Office; and as major program manager, Advanced Systems Program Office. RADM Fisher graduated from the U.S. Naval Academy with a B.S. in physics and from the Naval Postgraduate School with an M.S. in physics. He has been awarded the Distinguished Service Medal, Defense Superior Service Medal, Legion of Merit, Defense Meritorious Service Medal, and various other service medals and awards.

J. Daniel Stewart, *Vice Chair,* is the associate vice president for research at the University of Tennessee (UT). Since September 2004, in this role, he supports the UT executive vice president in overseeing research activities at its five campuses across the state as well as its research activities with strategic partners such as the Oak Ridge National Laboratory, which UT jointly manages with Battelle.

He also serves as an adjunct professor at the UT College of Business Administration. Before joining UT, Dr. Stewart served as the executive director of the Air Force Material Command at Wright-Patterson Air Force Base in Ohio. Dr. Stewart entered federal service in 1974 as a technology manager with the Air Force Rocket Propulsion Laboratory at Edwards Air Force Base in California. Before that, he served with The Aerospace Corporation, providing systems engineering and technical support to the Air Force's Space and Missile Systems Organization. In 1981, he transferred to Eglin AFB in Florida, where he held various mid-level to senior management positions involving development planning, acquisition, and test and evaluation. Before becoming the first executive director for the Air Force Material Command, he served as the executive director for the Air Armament Center, with responsibility for the development, test, acquisition, and sustainment of air-delivered weapon systems. He is a Stanford University Sloan Fellow and served details in Washington, D.C., supporting the congressionally mandated 1995 and 2005 base realignment and closure activities. Dr. Stewart holds B.S., M.S., and Ph.D. degrees in aerospace engineering from the Georgia Institute of Technology and a master's degree in management science from Stanford University. As a member of the Senior Executive Service, he was awarded two Distinguished and three Meritorious Presidential Rank Awards.

John A. Betti (NAE) is a retired Undersecretary of Defense for the Department of Defense (DOD). In that position at DOD, he had oversight responsibility for the Defense Research and Development and Acquisition process involving over $140 billion and 500,000 people. He also served as the National Armaments Director and the nation's representative to NATO and to the Four Powers (United States, France, Germany, and Great Britain) National Armament Directors' meetings. He was a member of the DOD Executive Committee and Defense Planning and Resources Board as well as chairman of the Defense Acquisition Board and the Defense Ethics Council. During his tenure with Ford Motor Company, his responsibilities included executive vice president of the Diversified Products Operation with responsibility for 12 businesses with total revenues of $13 billion (1988 dollars), 95,000 people, manufacturing plants in the United States, Canada, Mexico, Brazil, and six European countries, and joint ventures in Japan, Korea, and the United States. Six of the businesses engineered and manufactured electronic, electrical-mechanical, heating and air-conditioning, plastic, aluminum and iron castings and glass components for the automotive industry. The nonautomotive businesses were Ford Aerospace ($1.5 billion revenue, defense and aerospace), Ford New Holland ($3 billion revenue, tractor and farm implements), Rouge Steel ($1.5 billion revenue, steel), Philco Ford Brazil ($300 million revenue, television and VCR sets for Brazilian market), Ford Land Development (office buildings, shopping centers, hotels, and golf courses), and Hertz (automotive and equipment rentals). Mr. Betti also served as executive vice president for technical and operating staffs and was responsible for eight corporate staffs with worldwide responsibilities in manufacturing and engineering; purchasing and supply; quality; environmental and safety; research; design; marketing; and the Alpha Project (responsible for developing innovative manufacturing, engineering, product and business processes to ensure the company's competitiveness for products, quality, and cost into the twenty-first century).

Christopher L. Blake is the senior technical fellow for air system design and integration and the deputy to the vice president for enterprise process integration at Lockheed Martin Aeronautics (LM Aero). He is responsible for developing a solution to functional process integration to improve LM Aero program execution. He is also charged with improving the company's approaches to system design and integration and ensuring that they represent company and industry best practice. Earlier, Mr. Blake served as the leader and technical expert of several key systems engineering efforts, including Enterprise Standard Planning Package Development, F-35 Functional Baseline Development, and the Aerial Common Sensor LM Aero Independent Review Team. He has served as a systems engineering technical authority in support of many LM Aero key programs and is widely acknowledged as a company expert on systems engineering matters. From 1971 to 2004, Mr. Blake was a civilian with the DOD Aeronautical Systems Division at Headquarters Air Force Materiel Command, accumulating 33 years as a proven systems engineer and program manager in advanced development, acquisition, manufacturing, and sustainment of aircraft weapons systems. He was a senior executive for the Air Force's Aeronautical Systems Center and the Air Force Materiel Command. He held several key leadership positions: the associate director of plans and programs for the Air Force Materiel Command, primarily focused on base reallocation and closure (BRAC) planning, and director of engineering for the F-22A program and for the C-17 program. At LM Aero, Mr. Blake regularly represents the view of the customer to ensure LM Aero planning is in touch with customer expectations. A 1971 graduate of Wright State University with a B.S. degree in systems engineering, Mr. Blake earned a first master's degree in systems engineering in 1981, also from Wright State University, and a second master's degree in systems management in 1989 from Stanford University's Sloan Fellowship Program.

Claude M. Bolton, Jr., became the executive in residence for the Defense Acquisition University (DAU) on January 3, 2008. In this position, Mr. Bolton supports the DAU president, faculty, and students with strategic planning, course development, and mentoring. Mr. Bolton's primary focus is assisting the DAU president to achieve the congressional direction to recruit, retain, train, and educate the DOD acquisition workforce. Before that, Mr. Bolton served as the Assistant Secretary of the Army for Acquisition, Logistics and Technology (ASAALT), serving as the Army acquisition executive (AAE), the senior procurement executive, and the science advisor to the Secretary. Mr. Bolton was also the senior R&D official for the Army and had principal responsibility for all Army matters related to logistics.

In his role as ASAALT/AAE, Mr. Bolton led the Army's acquisition function and its acquisition management system. He oversaw the life-cycle management and sustainment of Army weapons systems and equipment, from R&D through test, evaluation, acquisition, logistics, fielding, and disposition. He oversaw the Army's program for the elimination of chemical weapons program and had oversight and executive authority over the Project and Contracting Office charged with Iraq reconstruction. Mr. Bolton was responsible for appointing, managing, and evaluating program executive officers as well as managing the Army Acquisition Corps and Army Acquisition Workforce. A veteran of more than 30 years of active military service, Mr. Bolton retired as a major general in the Air Force following a highly decorated career. In 1969, Mr. Bolton received his Air Force commission through the University of Nebraska's Air Force ROTC program. Mr. Bolton earned a bachelor's degree in electrical engineering from the University of Nebraska; a master's in management from Troy State University; and a master's in national security and strategic studies from the Naval War College. In July 2006, he was awarded a doctorate in science from Cranfield University in England. In May 2007, he was awarded an Honorary Doctor of Science degree from the University of Nebraska-Lincoln, his alma mater.

Allan V. Burman is president of Jefferson Solutions, a division of the Jefferson Consulting Group, LLC. Under his leadership, Solutions has provided change management services and/or acquisition reform training to many federal departments and agencies and other public service entities. He also has advised firms, congressional committees, and federal and state agencies on management and acquisition reform matters and regularly speaks to groups on related topics. Before joining The Jefferson Group in 1994, Dr. Burman had a lengthy career in the federal government, serving in policy positions in the White House's Office of Management and Budget (OMB) and in the Office of the Secretary of Defense. As the former administrator for federal procurement policy, he had the longest tenure of any administrator, serving in the Executive Office of the President under Presidents Reagan, Bush, and Clinton. As a member of the Senior Executive Service, Dr. Burman served in the early 1980s in OMB's National Security Division as chief of the Air Force Branch and was the first OMB Branch Chief to receive a Presidential Rank Award. Dr. Burman is a fellow of the National Academy of Public Administration, a principal of the Council for Excellence in Government, a director of the Procurement Round Table, a fellow and member of the board of advisors to the National Contract Management Association, and an honorary member of the National Defense Industrial Association. He is also an adjunct professor at the International Law Institute (ILI), a member of ILI's Procurement Advisory Board, and an adjunct professor at George Mason University.

John T. Dillard is a senior lecturer in the Graduate School of Business and Public Policy at the Naval Postgraduate School in Monterey, California. He joined the faculty in the fall of 2000 upon his retirement from the U.S. Army at the rank of Colonel. In the mid-1980s, he served as manager of Close Combat Systems at Picatinny Arsenal, New Jersey, and worked across a myriad of technologies and system concepts that have evolved into fielded products, such as the M-4 carbine, the 120-mm mortar, and the sniper weapon. Later, he was an assistant project manager for development of both the Army tactical missile system and the JAVELIN antitank weapon system at Redstone Arsenal, Alabama. These systems incorporate state-of-the-art technologies and are still in sustained production and fielding. Both have proved to be highly effective in combat. As product manager for the Joint Advanced Special Operations Radio System at Fort Monmouth, New Jersey, he successfully completed a 3-year prototyping effort on time and under budget. In 1998, he was appointed to head all DOD contract administration in the New York metropolitan area. His organization won the President's Quality Award, the federal government's equivalent to the Malcolm Baldrige Award for Quality, and held the title for 2 years. He has also served on the faculty of the U.S. Army War College and is now an adjunct professor for the University of California at Santa Cruz, educating Silicon Valley public and private industry professionals in project management. With extensive experience in system development and procurement, his research has primarily focused on defense acquisition policy changes and their implications. His work on centralized control of defense acquisition programs was used by the Defense Science Board in its 2005 report on management oversight in acquisition organizations and was again cited in the study *Beyond Goldwater-Nichols* on defense acquisition reform by the Center for Strategic and International Studies. His current research is oriented to spiral development and computational modeling of evolutionary acquisition.

Charles E. Franklin is the retired vice president of the Raytheon Company Evaluation Team. He was appointed to this position in September 2003. General Franklin joined Raytheon Company in 1998, and until his current assignment, served as president of its Integrated Defense Systems business. Before that, he served as vice president and general manager of the Raytheon Electronic Systems' Air and Missile Defense Systems business unit. Before joining Raytheon, he worked for Lockheed Martin-Sanders in Nashua, New Hampshire, as vice president for Programs and Mission Success and vice president for quality and mission success. Before Lockheed, General Franklin held the rank of lieutenant general, U.S. Air Force, and was commander of the Electronic Systems Center, Air Force Materiel Command, Hanscom Air Force Base. He was responsible for the Air Force's Center of Excellence, with more than 500 command and control, communications, and intelligence systems, handling more than $3 billion in programs annually. While at Hanscom, he also served as deputy commander, Tactical Systems, and deputy commander for the Joint Surveillance Target Attack Radar System. Before transferring to Hanscom Air Force Base, General Franklin held the position of program executive officer, Tactical and Airlift Programs, Air Force Executive

Office Organization, in Washington, D.C. His earlier assignments included system program director, advanced medium-range air-to-air missiles at Eglin Air Force Base, and commander, Rome Air Development Center, at Griffiss Air Force Base. General Franklin earned a bachelor's degree in mechanical engineering from the Georgia Institute of Technology and a master's degree in aeronautical-mechanical engineering from the Air Force Institute of Technology.

Charles L. Johnson II is vice president of Air Force Networks and Support Systems for Boeing Integrated Defense Systems. Before joining Boeing, he was a lieutenant general serving as both the PEO for C2ISR and as the Commander for the Electronic Systems Center (ESC), Hanscom Air Force Base. The center's mission is to acquire C2ISR systems and all cryptologic systems for the Air Force, Joint Services, and COCOMS. The organization comprises more than 12,000 people located at six sites throughout the United States. The men and women of the ESC manage more than $5 billion in programs annually in support of the Air Force and joint and coalition forces. General Johnson has performed the duties of an operations group commander and of a program director of the C-141 and C-17 system program offices, respectively. He was a command pilot with more than 4,000 hours in the B-52G, C-5A/B, C-17, C-141A/B, CH-3E, KC-135R, UH-1N, and F-I5E. Previously, the general was Commander, Oklahoma City Air Logistics Center, and Installation Commander, Tinker Air Force Base. General Johnson received a B.S. in civil engineering from the U.S. Air Force Academy and an M.S. in engineering administration and law from George Washington University. He also completed a business program on executive development and leadership at the University of Illinois; a program for senior executives in national and international study at the John F. Kennedy School of Government, Harvard University; and an acquisition professional development program, Certified Level III Program Management, at Defense Systems Management College.

Leslie F. Kenne is an independent consultant providing expertise and guidance to industry in the areas of program management, logistics, test and evaluation, and ethics compliance. She retired from the U.S. Air Force in 2003 after 32 years of military service. She serves on three corporate boards: Harris Corporation, Unisys, and now SRI International. General Kenne also serves on the National Research Council's Air Force Studies Board, which performs studies on topics selected by Air Force leadership. General Kenne graduated from Auburn University with a degree in aerospace engineering and entered the Air Force in 1971 as a distinguished graduate of the ROTC program. She served as a flight line maintenance officer in operations and attended the U.S. Air Force Test Pilot School in 1974. She served in various test and evaluation project management and supervisory positions in development, operational, and joint testing organizations during a 12-year period. General Kenne directed three major development programs while in the Air Force: the low-altitude navigation and targeting infrared system for night, the

F-16, and the Joint Strike Fighter. She also served as vice commander of the Aeronautical Systems Center at Wright-Patterson Air Force Base and the Sacramento Air Logistics Center at McClellan Air Force Base. Her last two positions before retiring were commander of the Electronic Systems Center at Hanscom Air Force Base and the deputy chief of staff for warfighting integration at the Pentagon.

Andrew P. Sage (NAE) is the founding dean emeritus of the School of Information Technology and Engineering and a professor at George Mason University. He is an elected fellow of the Institute of Electrical and Electronics Engineers, the American Association for the Advancement of Science, and the International Council on Systems Engineering. He is the editor of the John Wiley textbook series on systems engineering and management, the INCOSE Wiley journal *Systems Engineering*, and is coeditor of *Information, Knowledge, and Systems Management*. He edited *IEEE Transactions on Systems, Man, and Cybernetics* from January 1972 through December 1998 and also served for 2 years as president of the IEEE SMC Society. In 1994 he received the Donald G. Fink Prize from the IEEE and a Superior Public Service Award for his service on the CNA Corporation board of trustees from the U.S. Secretary of the Navy. In 2000, he received the Simon Ramo Medal from the IEEE in recognition of his contributions to systems engineering and an IEEE Third Millennium Medal. In 2002, he received an Eta Kappa Nu Eminent Membership Award and the INCOSE Pioneer Award. He was elected to the National Academy of Engineering in 2004 for contributions to the theory and practice of systems engineering and systems management. In 2007, he was elected as a charter member of the Omega Alpha systems engineering honor society. His interests include systems engineering and management efforts in a variety of application areas, including systems integration and architecting, reengineering, engineering economic systems, and sustainable development. Dr. Sage received a B.S.E.E. degree from the Citadel, a S.M.E.E. degree from MIT, and a Ph.D. from Purdue, the last in 1960. He has been a faculty member at several universities and in 1984 became First American Bank Professor of Information Technology and Engineering at George Mason University and the first dean of its School of Information Technology and Engineering.

Mark D. Schaeffer is executive director and chief systems engineer with ManTech SRS. Before that he was director of systems and software engineering (SSE), Office of the Under Secretary of Defense (Acquisition, Technology and Logistics). As director of SSE he served as the department's chief engineer responsible for policy, guidance, and education and training for systems engineering, development test and evaluation, risk management, reliability and maintainability, quality, production, and manufacturing. In addition, he provided engineering and development T&E support to approximately 180 DOD programs for fixed, rotary, and unmanned aircraft; land and sea systems; tactical missiles; communication; and command and control domains. He was the focal point for the department's efforts to revitalize systems engineering that included new DOD-wide SE policy, guidance and related education and training, establishment and leadership of a forum of chief engineers from the military services and agencies within DOD and restructuring of engineering assessments leading to major milestone review by the Defense Acquisition Board. Before SSE, he served as DARPA's chief operating officer and chief information officer with direct responsibility for day-to-day operations of all DARPA security (physical and programmatic), automated information systems, human resources, and facilities. He has held positions in the Navy Department, the Presidential Commission on Merchant Marine and Defense, and served on the congressional professional staff. Schaeffer received a B.S.M.E. from California State University and has completed executive education at MIT and Duke University. He was the DOD sponsor of the NDIA Systems Engineering Division; a DOD representative to INCOSE's Corporate Advisory Board; chairman, Stevens Institute of Technology, School of Systems and Enterprise Advisory Board; and chairman of the NATO Life Cycle Management Group and its Continuous Acquisition and Life Cycle Support Management Board.

George R. Schneiter is currently a consultant to Boeing Integrated Defense Systems and to the Institute for Defense Analyses, the Defense Science Board, and the Missile Defense Agency. Before that he held several positions with the Office of the Secretary of Defense, including director, Strategic and Tactical Systems, Office of the Under Secretary of Defense (Acquisition, Technology and Logistics), with responsibility for acquisition oversight of all DOD strategic and tactical systems (aircraft, missiles, ships, land vehicles, and electronic warfare) and related arms control implementation and compliance, developmental test and evaluation, and foreign comparative testing. He has also held positions with the Center for Naval Analyses as well as The Aerospace Corporation. Dr. Schneiter received a B.S.M.E., an M.S.M.E., and a Ph.D. in mechanical engineering from Purdue University. He received the Distinguished Executive Presidential Award; the Meritorious Executive Presidential Award (three times); the DOD Distinguished Civilian Service Medal (three times); Outstanding Mechanical Engineer, Purdue University School of Mechanical Engineering; Commander, National Order of Merit, Government of France; and is a fellow of the American Institute of Aeronautics and Astronautics.

Robert J. Skalamera is a professional consultant, working with selected clients in government, industry, and academia. His services include systems engineering, technical management, and risk management in major systems acquisition and sustainment. He was the recipient of the 2007 Lt Gen Thomas R. Ferguson Jr. Award for Systems Engineering Leadership of the National Defense Industrial Association. Before that, Mr. Skalamera was the Deputy Director of Systems and Software Engineering for the Office of the Secretary of Defense in Washington,

D.C. Mr. Skalamera is an accomplished leader in applied systems engineering with more than 35 years of experience in defense systems, research, development, integration, test, and sustainment. He is recognized nationally and internationally for expertise in systems engineering, risk management, enterprise development, total life-cycle management, program assessment, and related subordinate fields in both technical and technical management domains.

Richard Szafranski is a partner in Toffler Associates, where he was a founding member in 1996, a managing partner from 2000 through 2006, and is now responsible for managing its international growth. Toffler Associates has clients in Belgium, Germany, Italy, The Netherlands, South Korea, and Sweden, along with global firms headquartered in the United States. Toffler Associates is a strategic planning and advisory firm that helps organizations transition to a knowledge-based economy. His consulting portfolio includes working with CEOs, COOs, senior executives, and officials of U.S. and international government agencies in global defense, security, aerospace, manufacturing, communications, and services. He helps clients focus on competitive strategies, globalization, branding, new ventures, mergers and acquisitions, and organizational transformation and is a coach for many senior executives. Mr. Szafranski served as an independent director on the board of directors for SBS Technologies, Inc., an embedded computer company, from 2003 until its acquisition by General Electric, chairing the Management Development and Compensation Committee and serving on the Corporate Governance and Nominating Committee. As committee chair he introduced the company's first formal executive pay-for-performance program. He also served as an independent director on the board of directors for the Ceridian Corporation, a business services and benefits company, from 2006, taking the company from a publicly traded company to a private company in 2007. He has completed executive education at the Harvard Business School.

Randall S. Weidenheimer is currently director of program integration in the XonTech Systems operating unit of Northrop Grumman's Information Systems Sector. Colonel Weidenheimer began his career at the Air Force's space acquisition organization. He first worked on a program to measure the space environment and moved on to the Advanced Warning System program, which developed technology for the next-generation strategic warning infrared surveillance satellites. Upon graduation from the University of Arizona with a master's degree in physics, he began working at the Air Force Space Technology Center. He worked on a variety of acquisition, tracking, and pointing and directed energy weapon system technology efforts related to the Strategic Defense Initiative. He worked on both laser and neutral particle beam programs, moving from action officer to branch chief to division chief. In his last Air Force assignment, Colonel Weidenheimer served as the system program director and then as the Wing Commander, for the SBIRS Wing, which had responsibility for developing and fielding both the legacy

Defense Support Program satellites and the new SBIRS high satellites and payloads. Colonel Weidenheimer led a program office of 500 people, with a portfolio value of more than $40 billion and annual budgets in excess of $700 million. He also took a major defense acquisition program through the Nunn-McCurdy breach process. He retired from the Air Force in 2007 and joined Northrop Grumman shortly thereafter.

Rebecca A. Winston is president of Winston Strategic Management Consultants. She is a former chair of the board of the Project Management Institute (PMI) and is a fellow of the institute. An expert on project management in the fields of research and development, energy, environmental restoration, and national security, she is known throughout the United States and globally as a leader in program and project management, primarily on programs funded by the U.S. government. She is a graduate of the University of Nebraska's College of Law, has a B.S. in education from Nebraska Wesleyan University, and an M.S. in biology from Idaho State University. She is a licensed attorney in Iowa and Nebraska. Active in the PMI since 1993, Ms. Winston helped pioneer PMI's Specific Interest Groups (SIGs) in the 1990s, including the Project Earth and government SIGs, and was a founder and first co-chair of the Women in Project Management SIG. She served two terms on the PMI board of directors as director at large, secretary-treasurer, vice chair, and chair. She is also a member of the American Bar Association and the Association of Female Executives. Ms. Winston currently serves as a consultant to organizations such as the National Nuclear Security Administration, the Department of Energy, and the Department of Homeland Security on program and project management, project reviews, risk management, and vulnerability assessments. She has extensive experience in the areas of national defense and security and has worked closely with local, regional, and national officials, including Congress and the Pentagon.

Appendix B

Meetings and Speakers

MEETING 1
MAY 6-7, 2008
KECK CENTER OF THE NATIONAL ACADEMIES
WASHINGTON, D.C.

Perspectives on Program Reviews
Terry Jaggers, Deputy Assistant Secretary of the Air Force for Science, Technology and Engineering

Looming Specialty Program Assessments
Lt Col Ed Masterson, U.S. Air Force, Deputy Division Chief, Engineering and Technical Management Division, Office of the Deputy Assistant Secretary (Science, Technology and Engineering)

Program Support Review Initiatives
Kristen Baldwin, Office of the Under Secretary of Defense, Acquisition, Technology and Logistics, Department of Defense

How We Do Business at the Space and Missile Systems Center—Key Processes and Some of Our Best Practices
Steve Soderquist, Civilian, U.S. Air Force, Air Force Space Command, SMC/PID

Air Force Program Reviews
Michael Kalna, Civilian, Secretary of the Air Force/Acquisition Center of Excellence

Defense Acquisition Performance Assessment
Lt Gen Ron Kadish, U.S. Air Force (retired), Vice President and Partner, Booz Allen Hamilton

Air Force Materiel Command: New Focus, New Role
Brig Gen Janet Wolfenbarger, U.S. Air Force, Director of Intelligence and Requirements and Special Assistant for Command Transformation to the Commander, Air Force Materiel Command

Oversight, Command and Control (OC2)
Janet Hassan, Chief, Acquisition Chief Process Office

The IPA Process and Products
Jack Wormington, Vice President, Program Assessment, The Aerospace Corporation

MEETING 2
JUNE 17-18, 2008
NAS BUILDING
WASHINGTON, D.C.

Global Hawk Systems Engineering Focused Program Support Review
Gary Grann, Consultant (Alion Science and Technology)

Assessment of Air Force Acquisition Program Reviews
Larry Lawson, Executive Vice President and F-22 Program General Manager, Lockheed Martin Aeronautics

Integrating and Streamlining Acquisition Oversight
Judy Stokley, Deputy Air Force Program Executive Officer for Weapons and Executive Director, Air Armament Center

Air Force Systems Engineering Assessment Model (SEAM)
G. Richard Freeman, Technical Director, Air Force Center for Systems Engineering

GAO Perspective
Katherine Schinasi, Managing Director, Acquisition and Sourcing Management, Government Accountability Office

Structure Programs for Success
Maj Gen Jeff Riemer, Air Force Program Executive Officer for the F-22 Program, Office of the Assistant Secretary of the Air Force for Acquisition

Acquisition Management Framework
Kenneth Miller, Special Assistant for Acquisition Governance and Transparency to the Secretary of the Air Force

A Northrop Grumman Perspective
Glenn Lamartin, Vice President, Strategic Programs, Command and Control Systems Division, Northrop Grumman Mission Systems

Discussion with Paul Kaminski
Paul Kaminski, Chairman and CEO, Technovation, Inc.

MEETING 3
JULY 15-17, 2008
KECK CENTER OF THE NATIONAL ACADEMIES
WASHINGTON, D.C.

Overview of DOD Acquisition System DRAFT Policy Changes
Skip Hawthorne, Senior Systems Analyst, Office of the Under Secretary of Defense for Acquisition, Technology and Logistics

U.S. Air Force: A Strategic Workforce Assessment Overview
Andrew Green, Partner, A.T. Kearney, Inc., and Randy Garber, Partner, A.T. Kearney, Inc.

An Air Force Program Manager's Perspectives
Col Dave Madden, U.S. Air Force, Air Force Space Command, Commander, Global Positioning Systems Wing

A Lockheed Martin Perspective
Jeffrey K. Harris, Corporate Vice President and General Manager, Situational Awareness, Lockheed Martin

Discussion with General Speedy Martin
Gen Speedy Martin, U.S. Air Force (retired), Independent Consultant

NSS 03-01, Policy Discussion
RADM Victor C. See, Jr., U.S. Navy; Program Executive Officer, Space Systems, U.S. Navy; Director, Communications Systems Acquisition and Operations Directorate, National Reconnaissance Office; Commander, Space and Naval Warfare Systems Command Space Field Activity

Leadership in the 21st Century: The Challenge of Mega-Systems
GEN Joseph Yakovac, U.S. Army (retired), Consultant, JVM LLC, Senior Lecturer, Naval Postgraduate School

Discussion with Blaise Durante, Senior Executive Service
Blaise Durante, U.S. Air Force, Deputy Assistant Secretary for Acquisition Integration

SITE VISIT
JULY 31, 2008
KECK CENTER OF THE NATIONAL ACADEMIES
WASHINGTON, D.C.

Discussion with Dr. Gansler
Jacques S. Gansler, Professor, University of Maryland

MEETING 4
AUGUST 11-14, 2008
NAS BUILDING
WASHINGTON, D.C.

Discussion
Col Arnold H. Streland, Commander, TSAT Space Group, MILSATCOM Systems Wing, Space and Missile Systems Center

Discussion
Col Charles B. Sherwin, Jr., Commander, 654th Aeronautical Systems Squadron, Large Aircraft Infrared Countermeasures System Program Manager, Aeronautical Systems Center

Discussion
Col John R. Griggs, Deputy Director, Develop and Sustain Warfighting Systems, Air Force Materiel Command, HQ AFMC/CCO

Discussion
Col Dwyer Dennis, Commander, 551st Electronic Systems Wing, Electronic Systems Center

Discussion
Gen Bruce Carlson, Commander, Air Force Materiel Command

Appendix C

Related Studies

AFA (Air Force Association). 2009. *Fixing DOD Acquisition.* Available online at http://www.afa.org/edop/edop_8_01_08.asp. Last accessed on December 10, 2008.

Chenoweth, Mary E., Sarah Hunter, Brent Keltner, and David Adamson. 2004. *Organizational Policy Levers Can Affect Acquisition Reform Implementation in Air Force Repair Contracts,* MR-1711. Santa Monica, Calif.: RAND Corporation.

Day, Donald. 1982. *GAO Memorandum: MASAD/SDA Handbook, January 29, 1982.* Washington, D.C.: General Accounting Office (GAO).

DOD (Department of Defense). 2006. *Defense Acquisition Performance Assessment.* Washington, D.C.: Deputy Secretary of Defense.

GAO. 1993. *Waivers to Acquisition Workforce Training, Education, and Experience Requirements*, GAO/NIASD 93-128. Washington, D.C.: GAO.

GAO. 1993. *Implementation of the Defense Acquisition Workforce Act*, GAO/NIASD 93-129. Washington, D.C.: GAO.

GAO. 1996. *Best Practices: Commercial Quality Assurance Practices Offer Improvements for DOD,* GAO/NSIAD-96-162. Washington, D.C.: GAO.

GAO. 1998. *Best Practices: Successful Application to Weapons Acquisition Requires Changes in DOD's Environment,* GAO/NSIAD-98-56. Washington, D.C.: GAO.

GAO. 1998. *Implementation of Key Aspects of the Federal Acquisition Streamlining Act of 1994*, GAO/NSIAD-98-81. Washington, D.C.: GAO.

GAO. 1998. *Defense Acquisitions: Improved Program Outcomes Are Possible*, GAO/NSIAD-98-123. Washington, D.C.: GAO.

GAO. 1999. *Defense Acquisitions: Best Commercial Practices Can Improve Program Outcomes,* GAO/NSIAD-99-116. Washington, D.C.: GAO.

GAO. 1999. *Best Practices: Better Management of Technology Development Can Improve Weapons System Outcomes*, GAO/NSIAD-99-162. Washington, D.C.: GAO.

GAO. 1999. *Best Practices: DOD Training Can Do More to Help Weapons Systems Implement Best Practices*, GAO/NSIAD-99-206. Washington, D.C.: GAO.

GAO. 2000. *Defense Acquisitions: Employing Best Practices Can Shape Better Weapon System Decisions*, GAO/NSIAD-00-137. Washington, D.C.: GAO.

GAO. 2000. *Best Practices: A More Constructive Test Approach Is Key to Better Weapon Systems Outcomes*, GAO/NSIAD-00-199. Washington, D.C.: GAO.

GAO. 2001. *Best Practices: Better Matching of Needs and Resources Will Lead to Better Weapons System Outcomes*, GAO-01-288. Washington, D.C.: GAO.

GAO. 2002. *Defense Acquisitions: DOD Faces Challenges in Implementing Best Practices*, GAO-02-469T. Washington, D.C.: GAO.

GAO. 2002. *Best Practices: Capturing Design and Manufacturing Knowledge Early Improves Acquisition Outcomes,* GAO-02-701. Washington, D.C.: GAO.

GAO. 2002. *Defense Acquisitions: Factors Affecting Outcomes of Advanced Concept Technology Demonstration,* GAO-03-52. Washington, D.C.: GAO.

GAO. 2003. *Best Practices: Setting Requirements Differently Could Reduce Weapons Systems' Total Ownership Costs,* GAO-03-57. Washington, D.C.: GAO.

GAO. 2004. *Defense Acquisitions: Stronger Management Practices Are Needed to Improve DOD's Software-Intensive Weapon Acquisition,* GAO-04-393. Washington, D.C.: GAO.

GAO (Government Accountability Office[1]). November 2005. *Best Practices: Better Support of Weapon System Program Managers Needed to Improve Outcomes*, GAO-06-110. Washington, D.C.: GAO.

GAO. 2005. *DOD Acquisitions Outcomes: A Case for Change*, GAO-06-257T. Washington, D.C.: GAO.

GAO. 2006. *DOD Management Approach and Processes Not Well-Suited to Support Development of Global Information Grid*, GAO-06-211. Washington, D.C.: GAO.

GAO. 2006. *Best Practices: Stronger Practices Needed to Improve DOD Technology Transition Processes,* GAO-06-883. Washington, D.C.: GAO.

GAO. 2006. *Defense Acquisitions: Major Weapons Systems Continue to Experience Cost and Schedule Problems Under DOD's Revised Policy,* GAO-06-368. Washington, D.C.: GAO.

[1]Name was changed on July 7, 2004.

GAO. 2007. *Defense Acquisitions: Assessments of Selected Weapons Programs,* GAO-07-406SP. Washington, D.C.: GAO.

GAO. 2008. *Best Practices: Increased Focus on Requirements and Oversight Needed to Improve DOD's Acquisition Environment and Weapons Systems Quality,* GAO-08-294. Washington, D.C.: GAO.

GAO. 2008. *Defense Management: More Transparency Needed over the Financial and Human Capital Operations of the Joint Improvised Explosive Device Defeat Organization*, GAO-08-342. Washington, D.C.: GAO.

Hanks, Christopher H., Elliot I. Axelband, Shuna Lindsay, Mohammed Rehan Malik, and Brett D. Steele. 2005. *Reexamining Military Acquisition Reform: Are We There Yet?* MG-291. Santa Monica, Calif.: RAND Corporation.

Johnson, Stephen B. 2002. Bernard Schriever and the Scientific Vision. *Air Power History* 49(1):30-45.

Lorell, Mark A., Julia F. Lowell, and Obaid Younossi. 2006. *Evolutionary Acquisition: Implementation Challenges for Defense Space Programs*, MG-413. Santa Monica, Calif.: RAND Corporation.

McKinney, Ethan, Eugene Gholz, and Harvey M. Sapolsky. 1994. *Acquisition Reform— Lean 94-03*. Cambridge, Mass.: MIT Press.

Putnam, W.D. 1972. *The Evolution of the Air Force System Acquisition Management,* R-868-PR. Santa Monica, Calif.: RAND Corporation.

Pyles, Raymond A., and Hyman L. Shulman. 1995. *United States Air Force Fighter Support in Desert Storm*, MR-468-AF. Santa Monica, Calif.: RAND Corporation.

Stem, David, Michael Boito, and Obaid Younossi. 2006. *Systems Engineering and Program Management: Trends and Costs for Aircraft and Guided Weapons Programs*, MG-413. Santa Monica, Calif.: RAND Corporation.

Appendix D

Survey

To obtain data to help it accomplish the tasks set for it, the committee developed and conducted a survey of Air Force PMs. A copy of the survey in its entirety follows. The main purpose of the survey was to increase the number of people the committee talked to beyond the limited number the committee had contacted directly and to obtain additional quantitative data. The survey data were yet another form of information to augment what the committee members had learned from their research, interviews, and personal experience.

The committee employed a multistep process to produce the final survey:

1. The initial series of survey questions was developed from inputs from former government program managers and senior consultants with relevant DOD experience.
2. The draft set of survey questions was discussed with current Air Force PMs and senior functional support staff at one Air Force product center and the questions were refined.
3. A survey expert from NRC provided general guidelines on the conduct of the survey and data protection statements, on ensuring that the survey questions were objectively stated and structured to encourage survey takers to complete the survey, and on incorporating human factor considerations. The draft survey was improved using these guidelines. The NRC expert also provided advice on the approval process for surveys that are part of an NRC-administered study. This advice was followed to obtain NRC Institutional Review Board approval of the survey.
4. Survey format feedback and Air Force survey approval process information were also provided by the Air Force Manpower Agency Air Force

Survey Office, which is responsible for approval of surveys of Air Force personnel. This feedback was also used to refine the survey and to plan the schedule for survey approval.

5. A refined draft survey was provided to several current Air Force PMs of various grades asking them to check on the clarity and pertinence of the questions and to estimate how long it would take to complete the survey. This feedback was used to further revise the survey questions.
6. The draft survey was provided to the full committee membership individually for their review, and the comments were used to further refine and streamline the survey.
7. The near-final survey was reviewed by an NRC survey expert, who helped to clarify some questions and eliminate others in order to reduce the time necessary to complete the survey while preserving the potential to collect as useful data as possible.

After survey development had been completed, the final version of the survey was submitted in parallel to the NRC Institutional Review Board and to the Air Force formal survey approval process. Both the NRC and the Air Force approved the survey as it was submitted, and it received an official Air Force Survey Number (USAF SCN 08-045) and an expiration date (July 18, 2009).

The next step was distribution to the intended survey population—that is, to Air Force PMs. Several steps remained:

1. Each of the four Air Force product centers (the Aeronautical Systems Center, the Air Armament Center, the Electronic Systems Center, and the Space and Missile Systems Center) was tasked to provide a list of PMs employed there. The list was to include all the Acquisition Category (ACAT) I PMs at the center and a sampling of ACAT II and ACAT III PMs. Each provided the PM names and e-mail addresses, which were entered into the Web-based survey tool.
2. Each of the Program Executive Officers (PEOs) at the four product centers was notified of the survey and invited to participate, both because of their program management experience and to ensure they were aware of the questions being asked of their PMs. Next the PMs identified at the centers were invited by e-mail to take the survey and given the direct link to the Web-based survey tool.
3. To maximize participation in the survey, NRC staff used the Web-based tool to send reminders to the PMs. Also, based on the advice of the NRC survey expert, the committee's survey subgroup extended the survey window in conjunction with a final word of encouragement to the PMs who had not yet taken the survey. The survey data collection period was closed out on August 7, 2008.

4. When the survey was developed, criteria had been developed to gauge the success of the survey from the standpoint of how the data would be used. Four criteria were developed in consultation with the NRC expert: acceptable number of responses from each product center; acceptable percentage of response from the most senior (ACAT I) PMs; acceptable percentage of overall PM responses; and acceptable number of responses for each review that is specifically evaluated in report.

Once time for responding to the survey had run out, the results were reviewed against the success criteria and were judged to have met them (or not). The results were then reviewed by the data collection subgroup of the committee and specific reports were broken out, depending on the particular issue/topic of interest, and provided to the full committee for its use. The specific results of the survey that gave rise to a particular committee finding are discussed in the analysis section under the pertinent finding. The quantitative survey results (multiple choice) are shown next, for all the PMs who responded to the survey. The qualitative "written-in" (essay type) comments from the PMs have not been included in this appendix because several of them gave so much detail that their authors could be identified by people familiar with their programs, violating their privacy.

Because the committee recognized the challenges of constructing a survey and reporting its results (such as bias, demographics, and numbers of responses), it sought and received professional support from the NRC in, among other things, devising criteria to judge whether there was sufficient information to permit meaningful analysis. The committee also made promises to the Air Force regarding the use and anonymity of the data. Although there were more responses from ESC than from the other centers, when the data were partitioned in various ways the overall results show significant consistency across the four centers. With respect to potential survey bias, the committee considered the possibility that disgruntled (or "dissatisfied") PMs might be more likely to respond. However, responses from the PMs on the survey were thoughtful and balanced and the balance of positive and negative comments on the survey was very much in alignment with interview comments and discussions.

A final note about the use of survey data: the main use was to weigh and compare the positive and negative perceptions of program reviews and to suggest how the overall review process, as well as individual reviews, could be made more effective from the perspective of the PMs. The specific response breakout percentages for any individual question were seldom the focus—the relative balance was of more interest to the committee in most cases. No finding, conclusion, or recommendation of the committee is based solely on survey data; rather, they represent what the committee heard from all its sources.

SURVEY AND RESPONSE DATA

Intro Page

Purpose of Survey: The primary purpose of this survey is to collect information from AF program managers on how much time/effort is uniquely spent preparing for, participating in, and following up on tasks from higher level AF and OSD reviews, that would not otherwise have had to be spent for the purpose of good program management. The study committee is also interested in your assessments, both positive and negative, of the higher-level reviews you have participated in and any changes you would recommend. Collecting this information will help the committee to determine how to respond to the SAF/AQR-sponsored study objective to "Identify and evaluate options for streamlining, tailoring, integrating, or consolidating reviews of programs to increase the cost-effectiveness and to lessen workforce impact of the reviews as a whole." The entire Statement of Task for this study can be found at http://www8.nationalacademies.org/cp/projectview.aspx?key=48922 for reference.

Data Protection Statement: The detailed data collected in this survey will only be viewed by the three committee people assigned to collect the data (Randy Weidenheimer, Richard Szafranski, and Allan Burman), the chairman and vice-chairman of the study committee (Rand Fisher and Dan Stewart, respectively), and National Academy of Sciences professional staff members (Jim Garcia, Enita Williams, and Kamara Brown). Any reporting of the survey results will be at the summary level, with information related to specific people or programs removed. If the study group decides that any direct quotations from write-in sections of the survey would be useful to illustrate specific points, then the committee will attribute the quote to "an AF program manager" and remove all identifying information (and will confirm with the author that this has been done satisfactorily).

Survey Instructions, Structure, and Statement of Task

Instructions for Survey: Please complete all questions in the survey, marking questions "N/A" as appropriate. If necessary, you can save a partially completed survey and return to it later to answer the remaining questions.

Structure of Survey:

Section 1—Demographic Data Section—information on program manager and program

Section 2—Program Activity Overview Section—information on pertinent external reviews/reporting accomplished by the program

Section 3—Questions on Specific Reviews—information on time/effort spent on specific reviews/reporting accomplished by each program manager taking the survey

Section 4—Optional Section to Comment on Streamlining/Tailoring/Integrating/Consolidating Opportunities

Section 1—Demographic Data

Instructions: Please complete all questions in Section 1.

1.0 At which product center do you currently work?

	Response Count
AAC	9
ASC	17
ESC	41
SMC	12
Other	2
answered question	81
skipped question	2

1.1 How long have you been a program manager in your current position?

	Response Count
Less than 6 months	23
6 months but less than 1 year	8
1 year but less than 2 years	21
2 years but less than 3 years	19
3 years or more	12
answered question	83
skipped question	0

1.2 Including the time in your current job, how many total years experience do you have performing the function of a program manager (PM), whether this was your official title or not?

	Response Count
Less than 1 year	0
1 year but less than 3 years	7
3 years but less than 5 years	8
5 years but less than 7 years	14
7 years but less than 15 years	26
15 or more years	28
answered question	83
skipped question	0

1.3 How much acquisition experience do you have? For the purposes of this study, consider time spent in program offices as well as staff assignments that worked with the requirements definition process, the planning/programming/ budgeting process, or the acquisition policy/governance process. Also, include any time spent working in industry in an equivalent job to the government jobs identified above.

	Response Count
Less than 1 year	0
1 year but less than 3 years	5
3 years but less than 5 years	3
5 years but less than 7 years	7
7 years but less than 15 years	23
15 or more years	45
answered question	83
skipped question	0

1.3.a Of the total time stated in 1.3, how much of the time was spent in System Program Office (SPO)?

	Response Count
Less than 1 year	0
1 year but less than 3 years	8
3 years but less than 5 years	13
5 years but less than 7 years	12
7 years but less than 15 years	36
15 or more years	13
answered question	82
skipped question	1

1.4 How many hours do you work, on average, each week?

	Response Count
40-45 hours	7
46-50 hours	18
51-55 hours	20
56-60 hours	17
61-65 hours	8
66-70 hours	7
71 or more hours	3
Other	1
answered question	81
skipped question	2

1.5 Estimate the percentage of your time you spend on the following activities each week. The sum of all fields should total "100," including the write-in "Other" field at the bottom of the list. You must enter the number "0" in any activity field—including "Other"—that is not applicable. For other, please write in applicable examples in the field.

	Response Average
Personnel Activities (e.g., performance reports, hiring actions, recognition and promotion ceremonies, career counseling/mentoring, etc.)	12.60
Administrative Activities (e.g., Center and Wing staff meetings, facility issues, security and computer training, etc.)	13.43
Military Training (e.g., physical fitness, self-aid buddy care, LOAC training, exercise support, etc.)	6.34
Program Management—above-the-Wing level activities (e.g., verbal and written reporting to chain-of-command beyond the Wing, including PEO, HQ AF, and OSD reviews/reports)	19.56
Program Management—Wing-level and below activities (e.g., including gov't-only meetings as well as interactions with the contractors)	46.02
Other Activities	5.85
answered question	82
skipped question	1

1.6 On average each week, what percentage of your time is spent in direct contact with your contractors? This includes talking on the phone and on VTCs as well as in-person meetings.

		Response Count
Less than 1%		2
1% to 5%		6
6% to 10%		19
11% to 15%		16
16% to 20%		15
21% to 25%		4
26% to 30%		5
31% to 35%		2
36%-40%		6
More than 40%		3
N/A		1
Other		2
	answered question	81
	skipped question	2

1.7 Absent any other demands on your time, ideally how much time would you want to spend in contact with your contractors, on average, each week?

		Response Count
Less than 1%		0
1% to 5%		0
6% to 10%		9
11% to 15%		10
16% to 20%		13
21% to 25%		14
26% to 30%		5
31% to 35%		10
36%-40%		9
More than 40%		11
N/A		0
Other		0
	answered question	81
	skipped question	2

1.8 How many total people are in your program office? Include Mil/Civ/FFRDC/SETA.

	Response Count
Less than 5	1
5-10	2
11-20	11
21-40	13
41-60	11
61-80	7
81-100	2
101 or more	34
N/A	1
answered question	82
skipped question	1

1.9 In your opinion, during your tenure has the acquisition experience level of personnel in your program office increased, remained about the same, or decreased?

	Response Count
Increased	17
Remained about the same	29
Decreased	36
N/A	0
answered question	82
skipped question	1

1.10 What is the approximate annual budget of your program/portfolio?

	Response Count
Less than $25M	15
$25M to $50M	9
$51M to $75M	5
$76M to $100M	2
$101M to $150M	8
$151M to $300M	16
$301M to $500M	10
$501M to $700M	3
$700M or more	11
N/A	2
answered question	81
skipped question	2

1.11 What is the Acquisition Category (ACAT) of your current program? If you have more than one program, indicate the highest ACAT rating within your portfolio (with ACAT ID being the highest possible).

	Response Count
ACAT ID	25
ACAT IC	5
ACAT IAM	1
ACAT IAC	3
ACAT II	15
ACAT III	30
N/A	2
answered question	81
skipped question	2

1.12 During the period 1 Jan 06 to 30 May 08, what acquisition phase has your program been in? If your program transitioned between two (or more) phases during this period, mark all that apply. If you have more than one major program, please complete for each program, with #1 being largest dollar value program, then #2, then #3.

	Program #1	Program #2	Program #3	Response Count
Concept Refinement	5	6	4	14
Technology Development	13	12	6	29
System Development and Demonstration	36	25	11	54
Production and Deployment	42	16	8	49
Operations and Support	24	6	12	34
N/A	4	2	2	4
			answered question	81
			skipped question	2

1.13 One possible driver of the need for higher level reviews is to ensure the coordination of programs that have a significant amount of external interfaces. Please characterize the amount of external interfaces of the program with other efforts.

	Response Count
Stand-alone system with very little/minimal amount of external interfaces	10
Modest amount of external interfaces	28
Extensive amount of external interfaces	43
N/A	1
answered question	82
skipped question	1

Section 2—Program Activity Overview Section

If you have more than one program, then comment on the one with the highest Acquisition Category and, if more than one in that ACAT, comment on the program with highest total program cost.

NOTE: Reviews can have more than one purpose (e.g., approve a milestone, improve cost/schedule/technical performance, provide information, etc.). The questions below focus on specific purposes of reviews.

2.1 Which of the major program reviews/assessments has your program participated in during the period 1 Jan 06 to 30 May 08? Check all that apply.

	Response Count
Defense Acquisition Board (DAB) Milestone Review	10
Defense Space Acquisition Board (DSAB) Milestone Review	5
Defense Acquisition Board (DAB) Status Review	6
Defense Space Acquisition Board (DSAB) Status Review	2
Overarching Integrated Product Team (OIPT) Review	15
Technology Readiness Assessment (TRA)	21
Technology Maturity Assessment (TMA)	8
Independent Program Assessment (IPA)	23
Program Support Review (PSR)	11
Manufacturing Readiness Review (MRR)	7
Logistics Health Assessment (LHA)	5
System Engineering Assessment Model (SEAM)	9
Air Force Review Board (AFRB)	13
Other	29
answered question	57
skipped question	28

2.2 For each of these major program reviews/assessments that your program experienced, indicate your assessment of their impact on program performance (i.e., cost/schedule/technical performance accomplishment)? Please select a response for each review, indicate "N/A" for each review that is not applicable.

	Positive Impact	No Impact	Negative Impact	N/A	Response Count
Defense Acquisition Board (DAB) Milestone Review	7	5	1	35	48
Defense Space Acquisition Board (DSAB) Milestone Review	2	1	1	38	42
Defense Acquisition Board (DAB) Status Review	2	4	1	39	46
Defense Space Acquisition Board (DSAB) Status Review	2	0	2	39	43
Overarching Integrated Product Team (OIPT) Review	7	7	4	30	48
Technology Readiness Assessment (TRA)	8	13	2	27	50
Technology Maturity Assessment (TMA)	4	4	1	34	43
Independent Program Assessment (IPA)	12	5	4	27	48
Program Support Review (PSR)	4	5	2	33	44
Manufacturing Readiness Review (MRR)	5	3	0	34	42
Logistics Health Assessment (LHA)	3	2	2	35	42
System Engineering Assessment Model (SEAM)	3	7	2	33	45
Air Force Review Board (AFRB)	7	5	1	34	47
Other	9	8	3	23	43
			answered question		66
			skipped question		17

2.2.a Which single review had the greatest positive impact on program performance?

	Response Count
Defense Acquisition Board (DAB) Milestone Review	5
Defense Space Acquisition Board (DSAB) Milestone Review	0
Defense Acquisition Board (DAB) Status Review	1
Defense Space Acquisition Board (DSAB) Status Review	1
Overarching Integrated Product Team (OIPT) Review	3
Technology Readiness Assessment (TRA)	4
Technology Maturity Assessment (TMA)	1
Independent Program Assessment (IPA)	11
Program Support Review (PSR)	2
Manufacturing Readiness Review (MRR)	2
Logistics Health Assessment (LHA)	0
System Engineering Assessment Model (SEAM)	3
Air Force Review Board (AFRB)	3
Other	19
answered question	55
skipped question	28

2.3 Reference your highest rated review from question 2.2a above, why did this review have a positive impact? Check all that apply.

	Response Count
Subject matter experts provided valuable inputs on problems/issues	21
Senior leaders engaged to help resolve problems/issues	26
Visibility of the review focused contractor leadership attention on fixing problems prior to having to brief senior government leaders	16
Program office uncovered problems/issues as part of preparation for review	12
N/A	19
Other	7
answered question	64
skipped question	19

2.4 Higher level HQ AF/OSD reviews/assessments provide senior leadership information that is necessary for their understanding of program performance, to fulfill their oversight role. Please rate each of the reviews that your program experienced in terms of how effective you believe the structure/format of the review was at providing useful data to the senior AF and OSD leadership.

	Lots of Useful Data	Some Useful Data	Little Useful Data	No Useful Data	N/A	Response Count
Defense Acquisition Board (DAB) Milestone Review	4	10	2	0	26	42
Defense Space Acquisition Board (DSAB) Milestone Review	2	4	1	0	27	34
Defense Acquisition Board (DAB) Status Review	1	4	2	0	30	37
Defense Space Acquisition Board (DSAB) Status Review	0	1	2	0	31	34
Overarching Integrated Product Team (OIPT) Review	4	11	3	2	23	43
Technology Readiness Assessment (TRA)	8	9	6	2	19	44
Technology Maturity Assessment (TMA)	2	3	2	2	27	36
Independent Program Assessment (IPA)	9	9	0	2	21	41
Program Support Review (PSR)	1	7	3	2	26	39
Manufacturing Readiness Review (MRR)	2	4	3	2	27	38
Logistics Health Assessment (LHA)	0	3	3	2	28	36
System Engineering Assessment Model (SEAM)	2	3	3	3	26	37
Air Force Review Board (AFRB)	4	9	2	1	26	42
Other	4	9	2	1	22	38
					answered question	61
					skipped question	22

2.4.a Higher level HQ AF/OSD reviews/assessments provide senior leadership information that is necessary for their understanding of program performance, to fulfill their oversight role. Please rate each of the reviews that your program experienced in terms of how effective you believe the structure/format of the review was at providing useful data to the senior AF and OSD leadership.

	Lots of Useful Data	Some Useful Data	Little Useful Data	No Useful Data	N/A	Response Count
Defense Acquisition Board (DAB) Milestone Review	4	8	1	0	32	45
Defense Space Acquisition Board (DSAB) Milestone Review	1	4	1	1	31	38
Defense Acquisition Board (DAB) Status Review	1	5	1	1	33	41
Defense Space Acquisition Board (DSAB) Status Review	0	1	2	1	35	39
Overarching Integrated Product Team (OIPT) Review	5	6	6	1	27	45
Technology Readiness Assessment (TRA)	5	9	5	3	25	47
Technology Maturity Assessment (TMA)	1	4	2	2	31	40
Independent Program Assessment (IPA)	12	6	0	1	26	45
Program Support Review (PSR)	3	4	1	4	30	42
Manufacturing Readiness Review (MRR)	1	3	3	3	32	42
Logistics Health Assessment (LHA)	0	3	1	3	33	40
System Engineering Assessment Model (SEAM)	1	3	2	4	30	40
Air Force Review Board (AFRB)	5	7	1	1	31	45
Other	7	6	1	1	21	38
				answered question		61
				skipped question		22

2.5 From the list below, identify the three higher level HQ AF/OSD reviews/ reporting activities that you believe have the LEAST beneficial impact on program performance. One would be the LEAST beneficial, followed by two, then three.

	Least Beneficial	Second Least Beneficial	Third Least Beneficial	Response Count
Defense Acquisition Board (DAB) Milestone Review	1	1	1	3
Defense Space Acquisition Board (DSAB) Milestone Review	0	0	1	1
Defense Acquisition Board (DAB) Status Review	1	4	3	8
Defense Space Acquisition Board (DSAB) Status Review	0	0	1	1
Overarching Integrated Product Team (OIPT) Review	8	3	2	13
Technology Readiness Assessment (TRA)	4	4	3	11
Technology Maturity Assessment (TMA)	4	6	2	12
Independent Program Assessment (IPA)	1	3	3	7
Program Support Review (PSR)	7	4	5	16
Manufacturing Readiness Review (MRR)	2	4	2	8
Logistics Health Assessment (LHA)	3	1	4	8
System Engineering Assessment Model (SEAM)	7	4	6	17
Air Force Review Board (AFRB)	2	2	4	8
Other	4	1	0	5
			answered question	46
			skipped question	37

2.6 What could be done to improve the positive impact of these reviews? (Select as many as apply for the three reviews identified in question 2.5 above.)

	Least Beneficial	Second Least Beneficial	Third Least Beneficial	Response Count
Hold the review at a different time in the program lifecycle	3	1	1	5
Reduce frequency of reviews	9	5	5	14
Expand attendee list to include additional subject matter experts	3	2	0	4
Restrict attendee list to smaller group	13	8	6	15
Change charter	8	3	3	9
Combine with another review	15	18	15	23
Consolidate or reduce number of pre-reviews	17	9	11	21
Narrow focus of review	11	9	7	16
Shorten length of the meeting	10	7	6	11
Other	5	3	5	7

2.7 From what you know from any source, identify the program reviews that have the highest potential to be combined into a single useful review. Please select from the list of reviews below and use the write-in section to show the pairings/groupings (examples: Review M & Z; Review S, T, & Y; Report O & R).

	Response Count
Defense Acquisition Board (DAB) Milestone Review	7
Defense Space Acquisition Board (DSAB) Milestone Review	2
Defense Acquisition Board (DAB) Status Review	7
Defense Space Acquisition Board (DSAB) Status Review	4
Overarching Integrated Product Team (OIPT) Review	12
Technology Readiness Assessment (TRA)	24
Technology Maturity Assessment (TMA)	21
Independent Program Assessment (IPA)	11
Program Support Review (PSR)	12
Manufacturing Readiness Review (MRR)	9
Logistics Health Assessment (LHA)	12
System Engineering Assessment Model (SEAM)	12
Air Force Review Board (AFRB)	5
Other	3
answered question	35
skipped question	48

2.8 Which of the written/digital reporting mechanisms has your program used during the period 1 Jan 06 to 30 May 08? Check all that apply.

		Response Count
SMART		70
PoPS		68
SAR		18
DAES		21
Other		13
	answered question	71
	skipped question	12

2.9 Internal to your program office, do you use these written/digital reporting mechanisms in the day-to-day management of your program? Please mark Yes, No, or N/A for all reporting mechanisms.

	Yes	No	N/A	Response Count
SMART	22	49	1	72
PoPS	18	50	2	70
SAR	2	26	31	59
DAES	3	25	30	58
Other	2	12	18	32
		answered question		72
		skipped question		11

2.10 Have you received any feedback from the HQ AF or OSD level on the inputs you have provided for these written/digital reporting mechanisms? Please mark Yes, No, or N/A for all reporting mechanisms.

	Yes	No	N/A	Response Count
SMART	15	54	1	70
PoPS	11	54	2	67
SAR	7	17	34	58
DAES	8	16	33	57
Other	4	6	23	33
			answered question	71
			skipped question	12

2.11 How well do you think these written/digital reporting mechanisms do at providing an accurate and informative view of your program? Please rate each tool that you use on a scale of 1 to 5, with 1 indicating the tool does a very good job at providing an accurate and informative picture of your program and 5 indicating that the tool does a very poor job at providing an accurate and informative picture. Mark all reporting mechanisms.

	Very Good (1)	Good (2)	Acceptable (3)	Poor (4)	Very Poor (5)	N/A	Response Count
SMART	7	22	26	12	4	1	72
PoPS	1	13	29	21	6	1	71
SAR	3	6	11	3	1	37	61
DAES	3	6	10	5	0	36	60
Other	1	4	4	2	1	20	32
						answered question	72
						skipped question	11

2.12 For the following major reviews, please indicate your opinion about whether the documentation required by higher authorities to support each of the following reviews is Insufficient (In), About Right (AR), Excessive but Decreasing (E-D), Excessive and Stable (E-S), or Excessive and Increasing (E-I). Select N/A if you do not have experience with a particular review.

	In	AR	E-D	E-S	E-I	N/A	Response Count
Defense Acquisition Board (DAB) Milestone Review	1	9	2	5	9	31	57
Defense Space Acquisition Board (DSAB) Milestone Review	1	5	1	0	4	41	52
Defense Acquisition Board (DAB) Status Review	1	4	1	8	4	37	55
Defense Space Acquisition Board (DSAB) Status Review	1	5	1	2	1	41	51
Overarching Integrated Product Team (OIPT) Review	1	13	1	4	10	28	57
Technology Readiness Assessment (TRA)	1	18	1	5	8	24	57
Technology Maturity Assessment (TMA)	2	11	1	5	1	33	53
Independent Program Assessment (IPA)	0	11	3	6	3	31	54
Program Support Review (PSR)	0	13	2	3	3	32	53
Manufacturing Readiness Review (MRR)	2	9	2	3	0	36	52
Logistics Health Assessment (LHA)	0	6	2	3	4	36	51
System Engineering Assessment Model (SEAM)	1	4	3	3	8	31	50
Air Force Review Board (AFRB)	0	10	2	8	2	32	54
Other	0	4	0	2	1	20	27
						answered question	62
						skipped question	21

Section 3A—Questions on Specific Reviews

This section asks you to rate the **Most Helpful** and **Least Helpful** higher level reviews/assessments that your program has experienced sometime during the period 1 Jan 06 to 30 May 08.

Most Helpful Review: For the higher level review/assessment that you thought was **most helpful** to execution of your program (reference question 2.2 above), please answer the following questions:

3.1 Did the review occur at the most useful time in the schedule for program activities?

		Response Count
Yes		38
No		21
	answered question	59
	skipped question	24

3.2 Did the review result in the decision(s) necessary to allow the program to continue on schedule?

		Response Count
Yes		46
No		13
	answered question	59
	skipped question	24

3.2.a Was this result appropriate, given the situation?

		Response Count
Yes		53
No		5
	answered question	58
	skipped question	25

3.3 Did the right subject matter experts appropriate for this review attend the meeting?

		Response Count
Yes		44
No		7
Does not apply		9
	answered question	60
	skipped question	23

3.4 Did you receive timely guidance from this meeting?

		Response Count
Yes		35
No		16
Does not apply		9
	answered question	60
	skipped question	23

3.5 Did the review, or directed follow-up action, cause any change in the current execution of the program?

		Response Count
Yes		24
No		29
Does not apply		8
	answered question	61
	skipped question	22

3.6 Did the review, or directed follow-up action, cause any change in the future plans for the program?

		Response Count
Yes		31
No		22
Does not apply		7
	answered question	60
	skipped question	23

3.7 What percentage of program office senior leadership personnel (i.e., X of the Y senior SPO people, example 2 of 5 = 40%) were involved with the preparation for this review?

		Response Count
Less than 10%		3
10% to less than 20%		5
20% to less than 30%		8
30% to less than 40%		11
40% to less than 50%		5
50% or more		26
	answered question	58
	skipped question	25

3.8 Was the prime contractor asked to provide support for this meeting?

		Response Count
Yes		26
No		21
Does not apply		13
	answered question	60
	skipped question	23

3.8.a If yes, then did this support involve more than 20% of the contractor leadership personnel (i.e., X of the Y senior program people, example 2 of 5 = 40%)?

		Response Count
Yes		13
No		22
	answered question	35
	skipped question	48

Section 3A (cont)—Identifying Positive and Negative Impacts

This study is trying to determine if there are more efficient ways to perform the higher HQ review/reporting process, so the most useful data that can be collected involve the impact of higher HQ reviews/reports on program performance. One way to measure this is to identify positive and negative impacts of these reviews/reports on program performance.

The questions below seek to identify any specific examples of both positive impacts and negative impacts.

3.9 Can you identify any ways in which the program performance was improved (e.g., problems in the program that were resolved faster or discovered earlier) because of the attention provided by this higher level review/assessment? Write in:

Write-in responses withheld.

3.10 Can you identify any negative impacts on program performance (e.g., problems in the program that took longer to discover or longer to resolve) because of the time spent supporting this higher level review/assessment? Write in:

Write-in responses withheld.

Section 3A (cont)—Estimates of High Level Review Costs

Part of the purpose of this survey is to understand the costs associated with higher level reviews. These costs can take many forms, including "opportunity" costs (as addressed previously), but another aspect of these costs is the monetary value of the time the program office spends doing preparation, prebriefs, the review itself, coordination of meeting minutes/decision memoranda, and postreview follow-up.

The questions below ask for estimates of these costs in several variations.

3.11 How many hours of government personnel (mil & civ) were needed for the total support of this review?

Write-in responses withheld.

3.12 How many hours of FFRDC/SETA personnel were needed for the total support of this review?

Write-in responses withheld.

3.13 How many hours of prime/subcontractor personnel were needed for the total support of this review?

Write-in responses withheld.

3.14 It is possible that some/most of the hours spent to prepare for the review, as documented in questions 3.11, 3.12, and 3.13, would have been spent as part of good program management, even if there had not been a higher level review. Therefore, this question asks you to estimate how many hours were uniquely spent to prepare for this higher level review that would not have been spent for any other reason besides preparing for this review. Estimate the unique hours spent on the higher-level review:

Write-in responses withheld.

3.15 What could be done to improve the utility of this review? (Check all that apply)

	Response Count
Reduce frequency	7
Increase attendees	2
Decrease attendees	12
Change charter	6
Combine with another review	9
Consolidate or reduce number of pre-reviews	22
Narrow focus of review	11
Shorten length of the meeting	5
Change sequence in relation to other reviews	3
Better synchronize with other reviews	14
Nothing—review is fine as it is	10
Other	10
answered question	53
skipped question	30

Section 3B—Evaluating Least Helpful Higher Level Reviews/Assessments

Least Helpful Review—For the higher level review/assessment that you thought was **least** helpful (reference question 2.5 above), please answer the following questions:

3.16 Did the review occur at the most useful time in the schedule for program activities?

	Response Count
Yes	19
No	23
answered question	42
skipped question	41

3.17 Did the review result in the decision(s) necessary to allow the program to continue on schedule?

	Response Count
Yes	22
No	20
answered question	42
skipped question	41

3.17.a Given the information that was presented at the review, was this result appropriate?

	Response Count
Yes	31
No	10
answered question	41
skipped question	42

3.18 Did the subject matter experts appropriate for this review attend the meeting?

	Response Count
Yes	32
No	10
answered question	42
skipped question	41

3.19 Did you get timely guidance from this meeting?

	Response Count
Yes	16
No	26
answered question	42
skipped question	41

3.20 Did the review, or directed follow-up action, cause any change in the current execution of the program?

	Response Count
Yes	12
No	30
answered question	42
skipped question	41

3.21 Did the review, or directed follow-up action, cause any change in the future plans for the program?

		Response Count
Yes		14
No		28
	answered question	42
	skipped question	41

3.22 What percentage of senior program leadership personnel (i.e., X of the Y senior SPO people, example 2 of 5 = 40%) were involved with the preparation for this review?

		Response Count
Less than 10%		2
10% to less than 20%		9
20% to less than 30%		4
30% to less than 40%		5
40% to less than 50%		5
50% or more		15
	answered question	40
	skipped question	43

3.23 Was the prime contractor asked to provide support for this meeting?

		Response Count
Yes		23
No		11
N/A		8
	answered question	42
	skipped question	41

3.23.a If yes, then did this support involve more than 20% of the contractor leadership personnel (i.e., X of the Y senior program people, example 2 of 5 = 40%)?

		Response Count
Yes		12
No		16
	answered question	28
	skipped question	55

Section 3B (cont)—Identifying Positive and Negative Impacts

This study is trying to determine if there are more efficient ways to perform the higher HQ reviews/assessments, so the most useful data that can be collected involve the impact of higher HQ reviews/reports on program performance. One way to measure this is to identify positive and negative impacts of these reviews/reports on program performance.

The questions below seek to identify any specific examples of both positive impacts and negative impacts.

3.24 Can you identify any ways in which the program performance was improved (e.g., problems in the program that were resolved faster or discovered earlier) because of the attention provided by this higher level review/assessment? Write in:

Write-in responses withheld.

3.25 Can you identify any negative impacts on program performance (e.g., problems in the program that took longer to discover or longer to resolve) because of the time spent supporting this higher level review/assessment?

Write-in responses withheld.

Section 3B (cont)—Estimates of High Level Review Costs

Part of the purpose of this survey is to understand the costs associated with higher level reviews. These costs can take many forms, including "opportunity" costs (as addressed above), but another aspect of these costs is the monetary value of the time the program office spends doing preparation, prebriefs, the review itself, coordination of meeting minutes/decision memoranda, and postreview follow-up.

The questions below ask for estimates of these costs in several variations.

3.26 How many hours of government personnel (mil & civ) were needed for the total support of this review?

Write-in responses withheld.

3.27 How many hours of FFRDC/SETA personnel were needed for the total support of this review?

Write-in responses withheld.

3.28 How many hours of prime/subcontractor personnel were needed for the total support of this review?

Write-in responses withheld.

3.29 It is possible that some/most of the hours spent to prepare for the review, as documented in questions 3.26, 3.27, and 3.28, would have been spent as part of good program management, even if there had not been a higher level review. Therefore, this question asks you to estimate how many hours were uniquely spent to prepare for this higher level review that would not have been spent for any other reason besides preparing for this review. Estimate the unique hours spent on the higher-level review:

Write-in responses withheld.

3.30 What could be done to improve the utility of this review? (Select all those that apply.)

	Response Count
Reduce frequency	9
Increase attendees	1
Decrease attendees	12
Change charter	7
Combine with another review	13
Consolidate or reduce number of pre-reviews	12
Narrow focus of review	10
Shorten length of the meeting	5
Change sequence in relation to other reviews	3
Better synchronize with other reviews	9
Nothing—review is fine as it is	4
Other	5
answered question	39
skipped question	44

Section 4—Comments on Streamlining/Tailoring/Integrating/ Consolidating Opportunities

Please provide general comments as well as recommendations on streamlining, tailoring, integrating, and consolidating opportunities.

4.1 What opportunities for streamlining of higher-level reviews, not previously mentioned, would you recommend?

Write-in responses withheld.

4.2 What opportunities for tailoring of higher-level reviews, not previously mentioned, would you recommend?

Write-in responses withheld.

4.3 What opportunities for integrating of higher-level reviews, not previously mentioned, would you recommend?

Write-in responses withheld.

4.4 What opportunities for consolidating of higher-level reviews, not previously mentioned, would you recommend?

Write-in responses withheld.

4.5 Please provide any other comments you think would improve the ability of higher HQ review of AF acquisition programs to either enable senior leaders to perform their oversight role more effectively, and/or help the program being reviewed execute more effectively.

Write-in responses withheld.

Appendix E

Types of Reviews

ACQUISITION STRATEGY PANEL (ASP)

Air Force Instruction 63-101 describes the ASP as follows:[1]

3.7.2.1. The Acquisition Strategy Panel supports the SAE and other MDAs. ASPs are forums that evaluate proposed acquisition strategies to ensure all key viable alternatives have been considered and that the best recommendation is provided to the SAE and/or the program's MDA for approval.

3.7.2.2. The SAF/AQX-ACE is the SAE-chaired ASP process owner and secretariat for all ACAT I/IA and non-delegated ACAT II programs.

3.7.2.3. The field ACE offices are the ASP process owner and secretariat for all non-SAE chaired ACAT II and III PEO/DAO programs.

3.7.2.4. Information concerning SAE-chaired ASPs, such as the current draft template for briefings, can be found at the SAF/AQX-ACE ASP secretariat website. Additionally, similar information pertaining to non-SAE chaired ASPs can be found at each of the respective Field ACE websites which are accessible on the SAF ACE website.

3.7.2.5. Additional information regarding general ASP requirements can be found in AFFARS 5307.104-90, Acquisition Strategy Panels (ASPs).

---

[1]Air Force Instruction 63-101, April 17, 2009, Acquisition and Sustainment Life Cycle Management.

AD HOC REVIEWS

Ad hoc reviews may come from many sources, some with very short time horizons, some with longer. They can come from leadership at the Office of the Secretary of Defense (OSD) seeking up-to-date information before a Defense Acquisition Board (DAB), Defense Acquisition Executive Summary (DAES), or other event that triggers interest or concern. They can come from the need to provide up-to-date information to support an unscheduled event or circumstance. Ad hoc reviews can also be initiated by military service or agency leadership in much the same way and for the same reasons. Advocacy focused reviews (miscellaneous technical topics such as SW, T&E, production readiness, etc.) sponsored by subject matter experts (SMEs) are generally focused to support some higher-level reviews such as an overarching integrated product team (OIPT), DAB, or service/agency management review.

AIR FORCE AUDIT AGENCY (AFAA)

Air Force Mission Directive 17 describes the AFAA as follows:[2]

> The AFAA accomplishes the internal audit mission of the United States Air Force. The AFAA provides timely, value-added audit services to all management levels. These services focus on independent, objective, and quality audits that include reviewing and promoting the economy, effectiveness, and efficiency of operations; assessing and improving Air Force fiduciary stewardship and the accuracy of financial reporting; and evaluating programs and activities and assisting management in achieving intended results.

AIR FORCE REQUIREMENTS FOR OPERATIONAL CAPABILITIES COUNCIL (AFROCC)

Air Force Instruction 10-601 describes the AFROCC as follows:[3]

> The AFROCC, an instrument of the CSAF and Secretary of the Air Force (SECAF), reviews, validates, and recommends approval of all Air Force capabilities-based requirements. The AFROCC ensures Air Force capabilities-based requirements documentation is prepared in accordance with Air Force and Joint Staff guidance, complies with established standards, and accurately articulates valid Air Force capabilities-based requirements. The AFROCC reviews Air Force FSA study plans directed by JCDs, AFCDs and for initiatives forecast to become ACAT I programs. For follow-on capabilities-based requirements documents, the AFROCC validates all Air Force-developed AoA Study Plans, interim status (when appropriate), and

[2]Air Force Mission Directive 17, November 13, 2002, Air Force Audit Agency (AFAA).

[3]Air Force Instruction 10-601, July 31, 2006, Capabilities-Based Requirements Development.

final results. It is chaired by AF/A5R and is composed of MAJCOM requirements principals, Secretariat, and HQ Air Force representatives.

AIR FORCE REVIEW BOARD (AFRB)

Air Force Instruction 63-101 describes the AFRB as follows:[4]

3.7.1.1. AF Review Boards are forums chaired by the SAE for conducting major decision reviews (in- or out-of-cycle), as well as making and documenting major milestone decisions. AFRBs are not conducted for services or space programs.

3.7.1.2. SAF/AQX-ACE is the AFRB process owner and secretariat.

3.7.1.3. The AFRB process is required for all ACAT IC, ACAT IAC, non-delegated ACAT II programs and special interest programs. The PEO may recommend what type of AFRB is necessary: full, mini (tailored attendance), or paper. A template and more information can be found at the SAF/AQX-ACE website.

3.7.1.4. For ACAT ID and ACAT IAMs, AFRBs are used to develop the AF corporate consensus prior to an OSD Defense Acquisition Board (DAB) (pre-DAB within AF) or Information Technology Acquisition Board (ITAB). The AFRB should be conducted no later than two weeks prior to last OSD Overarching Integrated Product Team (OIPT). The SAE determines if an ACAT ID or ACAT IAM program requires an AFRB.

3.7.1.5. PEOs and DAOs execute a tailored AFRB process for delegated ACAT II and ACAT III programs.

AIR FORCE SYSTEMS ENGINEERING ASSESSMENT MODEL (SEAM)

The Air Force Center for Systems Engineering describes the AF SEAM as follows:[5]

AF SEAM defines ten AF standard SE process areas, lists associated goals under each process area and provides associated specific and generic practices. Many of the best practices contained in AF SEAM were derived from various Software Engineering Institute (SEI)/Carnegie Mellon, Capability Maturity Model Integration® (CMMI®) products. Additionally, various international and industry standards, Department of Defense publications and development team

[4]Air Force Instruction 63-101, April 17, 2009, Acquisition and Sustainment Life Cycle Management.

[5]Air Force Systems Engineering Assessment Model (AF SEAM) Management Guide, Version 1, August 1, 2008. Available at http://www.afit.edu/cse/docs/AF%20SEAM%20Management%20Guide%20(Aug%202008).pdf. Last accessed May 4, 2009.

members' expert knowledge significantly contributed to the material contained in this model. It is essential to note that AF SEAM is a process assessment tool which is designed to assess the presence of needed SE processes as a "leading indicator" to subsequent delivery success. While the tool assesses the existence of SE process work products (i.e. CONOPS, plans, technical documents, etc) it does not assess the outcomes delivered to the customer. The model concentrates on "what" SE processes must be in place which, when properly executed, increase the likelihood customer needs will be satisfied. This is due to the fact that the quality of a System or Product is highly influenced by the quality of the process used to develop and maintain it.

ALTERNATIVE SYSTEM REVIEW (ASR)

The Defense Acquisition Guidebook describes the ASR as follows:[6]

The ASR is a multi-disciplined technical review to ensure that the resulting set of requirements agrees with the customers' needs and expectations and that the system under review can proceed into the Technology Development phase. The ASR should be complete prior to Milestone A. Generally this review assesses the alternative systems that have been evaluated during the Concept Refinement phase, and ensures that the preferred system alternative is cost effective, affordable, operationally effective and suitable, and can be developed to provide a timely solution to a need at an acceptable level of risk. Of critical importance to this review is the understanding of available system concepts to meet the capabilities described in the Initial Capabilities Document and the affordability, operational effectiveness, and technology risks inherent in each alternative concept. Depending on the overall acquisition strategy, one or more preferred solutions may carry forward into the Technology Development phase.

By reviewing alternative system concepts, the ASR helps ensure that sufficient effort has been given to conducting trade studies that consider and incorporate alternative system designs that may more effectively and efficiently meet the defined capabilities. A successful review is predicated on the IPT's determination that the operational capabilities, preferred solution(s), available technologies, and program resources (funding, schedule, staffing, and processes) form a satisfactory basis for proceeding into the Technology Development phase. The program manager should tailor the review to the technical scope and risk of the system, and address the ASR in the Systems Engineering Plan.

[6]Defense Acquisition Guidebook, Section 4.3.1.4.2, Alternative System Review (ASR). Available online at https://akss.dau.mil/dag/GuideBook/IG_c4.3.1.4.2.asp. Last accessed May 4, 2009.

CONFIGURATION STEERING BOARD (CSB)

Department of Defense Instruction 5000.02 describes the CSB as follows:[7]

The Acquisition Executive of each DoD Component shall establish and chair a CSB with broad executive membership including senior representatives from the Office of the USD(AT&L) and the Joint Staff. Additional executive members shall include representatives from the office of the chief of staff of the Armed Force concerned, other Armed Forces representatives where appropriate, the military deputy to the CAE and the Program Executive Officer (PEO) (section 814 of P.L. 110-417, Reference (w)).

(1) The CSB shall meet at least annually to review all requirements changes and any significant technical configuration changes for ACAT I and IA programs in development that have the potential to result in cost and schedule impacts to the program. Such changes will generally be rejected, deferring them to future blocks or increments. Changes shall not be approved unless funds are identified and schedule impacts mitigated.

(2) The PM, in consultation with the PEO, shall, on a roughly annual basis, identify and propose a set of descoping options, with supporting rationale addressing operational implications, to the CSB that reduce program cost or moderate requirements. The CSB shall recommend to the MDA (if an ACAT ID or IAM program) which of these options should be implemented. Final decisions on descoping option implementation shall be coordinated with the Joint Staff and military department requirements officials.

CRITICAL DESIGN REVIEW (CDR)

The Defense Acquisition Guidebook describes the CDR as follows:[8]

The CDR is a multi-disciplined technical review to ensure that the system under review can proceed into system fabrication, demonstration, and test; and can meet the stated performance requirements within cost (program budget), schedule (program schedule), risk, and other system constraints. Generally this review assesses the system final design as captured in product specifications for each configuration item in the system (product baseline), and ensures that each product in the product baseline has been captured in the detailed design documentation. Product specifications for hardware enable the fabrication of configuration items, and may include production drawings. Product specifications for software (e.g., Software Design Documents) enable coding of a Computer Software Configura-

[7]Department of Defense Instruction 5000.02, Operation of the Defense Acquisition System, December 8, 2008.

[8]Defense Acquisition Guidebook, Section 4.3.3.4.5, Critical Design Review (CDR). Available online at https://akss.dau.mil/dag/GuideBook/IG_c4.3.3.4.5.asp. Last accessed May 4, 2009.

tion Item. Configuration items may consist of hardware and software elements, and include items such as airframe, avionics, weapons, crew systems, engines, trainers/training, etc.

DEFENSE ACQUISITION BOARD (DAB)

The Defense Acquisition Guidebook describes the DAB as follows:[9]

The Defense Acquisition Board advises the USD(AT&L)/DAE on critical acquisition decisions. The USD(AT&L) chairs the Defense Acquisition Board, and the Vice Chairman of the Joint Chiefs of Staff serves as co-chair. Defense Acquisition Board members are the following executives: Under Secretary of Defense (Comptroller); Under Secretary of Defense (Policy); Under Secretary of Defense (Personnel & Readiness); Under Secretary of Defense (Intelligence); Assistant Secretary of Defense for Networks and Information Integration/DoD Chief Information Officer; Director, Operational Test & Evaluation; Chairman, Program Analysis and Evaluation; the Secretaries of the Army, the Navy, and the Air Force; and the Director, Acquisition Resources & Analysis (as the DAB Executive Secretary). Defense Acquisition Board advisors include the Principal Deputy USD(AT&L); the Deputy Under Secretary of Defense (Logistics & Material Readiness); the Director, Defense Research & Engineering; the relevant OIPT Leader(s); the Program Executive Officer; the Program Manager; the Chairmen, Cost Analysis Improvement Group; the Director, Defense Procurement and Acquisition Policy; DoD General Counsel; the Deputy Under Secretary of Defense (Industrial Policy); the DoD Component Acquisition Executives; Commander, United States Joint Forces Command; and the Chair, Functional Capabilities Board(s). The USD(AT&L)/DAE may ask other department officials to participate in reviews, as required.

DEFENSE ACQUISITION EXECUTIVE SUMMARY (DAES)

The Defense Acquisition Guidebook describes the DAES as follows:[10]

The DAES is a multi-part document, reporting program information and assessments; program manager, Program Executive Officer, CAE comments; and cost and funding data. The DAES provides an early-warning report to USD(AT&L) and ASD(NII). The DAES describes actual program problems, warns of potential program problems, and describes mitigating actions taken or planned. The program manager may obtain permission from USD(AT&L) or ASD(NII) to tailor DAES content. At minimum, the DAES should report program assessments

[9]Defense Acquisition Guidebook, Section 10.2.1, Defense Acquisition Board Review. Available online at https://akss.dau.mil/DAG/Guidebook/IG_c10.2.asp#1021. Last accessed May 4, 2009.

[10]Defense Acquisition Guidebook, Section 10.9.4, Defense Acquisition Executive Summary (DAES), Available online at https://akss.dau.mil/DAG/GuideBook/IG_c10.9.4.asp. Last accessed May 4, 2009.

(including interoperability), unit costs (10 U.S.C. 2433), and current estimates. It should also report the status of exit criteria and vulnerability assessments (31 U.S.C. 9106).

The DAES should present total costs and quantities for all years, as projected, through the end of the current acquisition phase. In keeping with the concept of total program reporting, the DAES should present best estimates for costs beyond the FYDP, if the FYDP does not otherwise identify those costs. (The total program concept refers to system acquisition activities from Program Initiation through Production and Deployment.) The DAES should also report approved program funding for programs that are subsystems to platforms and whose procurement is reported in the platform budget line.

DESIGN READINESS REVIEW (DRR)

Department of Defense Instruction 5000.2 described the DRR as follows:[11]

> The Design Readiness Review during SDD provides an opportunity for mid-phase assessment of design maturity as evidenced by measures such as the number of subsystem and system design reviews successfully completed; the percentage of drawings completed; planned corrective actions to hardware/software deficiencies; adequate development testing; an assessment of environment, safety and occupational health risks; a completed failure modes and effects analysis; the identification of key system characteristics and critical manufacturing processes; an estimate of system reliability based on demonstrated reliability rates; etc. Successful completion of the Design Readiness Review ends System Integration and continues the SDD phase into the System Demonstration effort. MDAs may, consistent with the intent of this paragraph, determine the form and content of the review.

DOD INSPECTOR GENERAL (DOD IG)

Department of Defense Directive 5106.01 describes the DOD IG as follows:[12]

> The Office of the Inspector General of the Department of Defense was established by Congress in the Defense Authorization Act for Fiscal Year 1983, Public Law (Pub. L.) 97-252, which is codified at Reference (c), as an independent and objective unit within the Department of Defense to conduct and supervise audits and investigations relating to the programs and operations of the Department of

[11] Department of Defense Instruction 5000.2, Operation of the Defense Acquisition System, May 12, 2003. (This instruction has since been superseded by DODI 5000.02, Operation of the Defense Acquisition System, December 8, 2008.)

[12] Department of Defense Directive 5106.01, Inspector General of the Department of Defense, April 13, 2006.

Defense. In support of the mission of the Department of Defense, the Inspector General performs the duties, has the responsibilities, and exercises the powers specified in Reference (c). [Appendix 3 of title 5, United States Code, "Inspector General Act of 1978," as amended]

FUNCTIONAL CONFIGURATION AUDIT (FCA)

Military Handbook 61A describes the FCA as follows:[13]

The Functional Configuration Audit (FCA) is used to verify that the actual performance of the CI meets the requirements stated in its performance specification and to certify that the CI has met those requirements. For systems, the FCA is used to verify that the actual performance of the system meets the requirements stated in the system performance specification. In some cases, especially for very large, complex CIs and systems, the audits may be accomplished in increments. Each increment may address a specific functional area of the system/CI and will document any discrepancies that are found in the performance capabilities of that increment. After all of the increments have been completed, a final (summary) FCA may be held to address the status of all of the action items that have been identified by the incremental meetings and to document the status of the FCA for the system or CI in the minutes and certifications. In this way, the audit is effectively accomplished with a minimum of complications.

GOVERNMENT ACCOUNTABILITY OFFICE (GAO)

In its performance plan for fiscal year 2009, the GAO describes itself as follows:[14]

GAO is an independent, nonpartisan, professional services agency in the legislative branch of the federal government. Commonly known as the "audit and investigative arm of the Congress" or the "congressional watchdog," we examine how taxpayer dollars are spent and advise lawmakers and agency heads on ways to make government work better.

Our mission is to support the Congress in meeting its constitutional responsibilities and to help improve the performance and ensure the accountability of the federal government for the benefit of the American people. We accomplish our mission by providing reliable information and informed analysis to the Congress, to federal agencies, and to the public, and we recommend improvements, when appropriate, on a wide variety of issues.

---

[13]Military Handbook 61A(SE), Configuration Management Guidance, February 7, 2001.

[14]Performance Plan for Fiscal Year 2009: Mission, Performance Plans, Resources and Strategies, GAO-08-507SP, February 19, 2008. Available online at http://www.gao.gov/htext/d08507sp.html. Last accessed May 5, 2009.

INDEPENDENT PROGRAM ASSESSMENT (IPA)

National Security Space Acquisition Policy 03-01 describes the IPA as follows:[15]

> An IPA is a focused, short duration "peer review" activity that typically runs from two to five weeks in duration depending on the program's complexity. The core members of an IPAT are assigned to work the assessment full-time for the IPAT leader, who is responsible for the final recommendation to the MDA. The IPA activity is usually conducted at the program office locale and/or the contractor facility to facilitate easy, ready access to the system experts, the data, and the equipment under review. While the IPAT may discuss issues with various elements in conducting the assessment, the assessment is not a consensus process. Rather, it produces an unbiased, structured, independent evaluation of the proposed space acquisition activity in order to provide the DoD Space MDA an overview of how well the SPD/PM has addressed problematic issues and to identify areas of concern or potential risk. The IPA will also report on vulnerability, mitigation and protection measures addressed by the program. The IPA also compares program accomplishment with program objectives and with previous DoD Space MDA direction, guidance, decisions, and/or Presidential or Congressionally directed actions.

INTEGRATED BASELINE REVIEW (IBR)

The Defense Acquisition Guidebook describes the IBR as follows:[16]

> An IBR is a joint assessment of the Performance Measurement Baseline (PMB) conducted by the government program manager and the contractor. The IBR is not a one-time event. It is a process, and the plan should be continually evaluated as changes to the baseline are made (modifications, restructuring, etc.). IBRs should be used as necessary throughout the life of a project to facilitate and maintain mutual understanding of:
>
> - The scope of the PMB consistent with authorizing documents;
> - Management control processes;
> - Risks in the PMB associated with cost, schedules, and resources; and
> - Corrective actions where necessary.

---

[15]National Security Space Acquisition Policy 03-01, Guidance for DoD Space System Acquisition Process, December 27, 2004.

[16]Defense Acquisition Guidebook, Section 11.3.3.3, Integrated Baseline Reviews (IBRs). Available online at https://akss.dau.mil/dag/DoD5000.asp?view=document&rf=GuideBook\IG_c11.3.1.3.asp. Last accessed May 4, 2009.

INTEGRATING INTEGRATED PRODUCT TEAM (IIPT)

The Defense Acquisition Guidebook mentions the IIPT as follows:[17]

> IPTs are an integral part of the Defense acquisition oversight and review process. For Acquisition Category ID and IAM programs, there are generally two levels of IPT: the Overarching Integrated Product Team and the Working-level Integrated Product Team(s). Each program should have an OIPT and at least one WIPT. WIPTs should focus on a particular topic such as cost/performance, test, or contracting. An Integrating Integrated Product Team (IIPT), which is itself a WIPT, should coordinate WIPT efforts and cover all topics not otherwise assigned to another IPT. IPT participation is the primary way for any organization to participate in the acquisition program.

JOINT REQUIREMENTS OVERSIGHT COUNCIL (JROC)

Chairman of the Joint Chiefs of Staff Instruction 5123.01D describes the JROC as follows:[18]

> a. JROC Mission. Title 10, United States Code (USC), section 181, directed the Secretary of Defense to establish the JROC. In addition to other matters assigned to it by the President or Secretary of Defense, the JROC shall:
>
> (1) Assist the Chairman in identifying and assessing the priority of joint military capabilities (including existing systems and equipment) to meet the national military and defense strategies.
>
> (2) Assist the Chairman in considering alternatives to any acquisition program that has been identified to meet military capabilities by evaluating the cost, schedule, and performance criteria of the program and of the identified alternatives.
>
> (3) As part of its mission to assist the Chairman in assigning joint priority among existing and future programs meeting valid capabilities, ensure that the assignment of such priorities conforms to and reflects resource levels projected by the Secretary of Defense through the JPG.
>
> b. JROC Membership. The Chairman is the chairman of the JROC. The functions of the JROC Chairman are delegated to the Vice Chairman of the Joint Chiefs of Staff. Other members of the JROC are officers in the grade of general or admiral from the Army, Navy, Air Force, and Marine Corps. Service representatives are recommended by their military department secretary and approved by the Chairman after consultation with the Secretary of Defense.

[17]Defense Acquisition Guidebook, Section 10.3, Role of Integrated Product Teams (IPTs). Available online at https://akss.dau.mil/DAG/Guidebook/IG_c10.3.asp. Last accessed May 5, 2009.

[18]Chairman of the Joint Chiefs of Staff Instruction 5123.01D, Charter of the Joint Requirements Oversight Council, August 1, 2007.

MANUFACTURING READINESS ASSESSMENT (MRA)

The draft Department of Defense Manufacturing Readiness Assessment (MRA) Deskbook describes the MRA as follows:[19]

> A Manufacturing Readiness Assessment (MRA), for the purposes of this document, is a structured evaluation of a technology, component, manufacturing process, weapon system or subsystem using the MRL definitions as a standard. It is performed to:
>
> - Define current level of manufacturing maturity
> - Identify maturity shortfalls and associated risks
> - Provide the basis for manufacturing maturation and risk management (planning, identification, analysis, mitigation, implementation, and tracking)

OPERATIONAL TEST READINESS REVIEW (OTRR)

The Defense Acquisition Guidebook describes the OTRR as follows:[20]

> The OTRR is a multi-disciplined product and process assessment to ensure that the "production configuration" system can proceed into Initial Operational Test and Evaluation with a high probability of successfully completing the operational testing. Successful performance during operational test generally indicates that the system is suitable and effective for service introduction. The Full Rate Production Decision may hinge on this successful determination. The understanding of available system performance to meet the Capability Production Document is important to the OTRR. The OTRR is complete when the Service Acquisition Executive evaluates and determines materiel system readiness for Initial Operational Test and Evaluation.

OVERARCHING INTEGRATED PRODUCT TEAM (OIPT)

The Defense Acquisition Guidebook describes the OIPT as follows:[21]

> All Acquisition Category ID and IAM programs will have an OIPT to provide assistance, oversight, and review as the program proceeds through its acquisition

---

[19]Manufacturing Readiness Assessment (MRA) Deskbook [draft], May 29, 2008. Available online at https://acc.dau.mil/GetAttachment.aspx?id=182129&pname=file&aid=34013&lang=en-US. Last accessed May 5, 2009.

[20]Defense Acquisition Guidebook, Section 4.3.4.4.2, Operational Test Readiness Review (OTRR). Available online at https://akss.dau.mil/dag/GuideBook/IG_c4.3.4.4.2.asp. Last accessed May 4, 2009.

[21]Defense Acquisition Guidebook, Section 10.3.1, Overarching IPT (OIPT) Procedures and Assessment. Available online at https://akss.dau.mil/dag/Guidebook/IG_c10.3.1.asp. Last accessed May 4, 2009.

life cycle. An appropriate official within OSD, typically the Director, Defense Systems or the Deputy to the Assistant Secretary of Defense for Networks and Information Integration (ASD(NII)) for Command, Control, Communications, Intelligence, Surveillance and Reconnaisance (C3ISR) [sic] and Information Technology (IT) Acquisition, will lead the OIPT for Acquisition Category ID programs. The Deputy to the ASD(NII) for C3ISR and IT Acquisition also leads the OIPT for Acquisition Category IAM programs. The OIPT for Acquisition Category IAM programs is called the NII OIPT. OIPTs should include the Program Manager, Program Executive Officer, DoD Component Staff, Joint Staff, and OSD staff involved in oversight and review of the particular Acquisition Category ID or IAM program. Other OIPTs, specifically those for Chem Bio and Space, will be lead [sic] and directed by similar executives.

The OIPT should form upon departmental intention to start an acquisition program. The OIPT charters the Integrating Integrated Product Team and Working-level Integrated Product Teams. The OIPT should consider the recommendations of the Integrating Integrated Product Team regarding the appropriate milestone for program initiation and the minimum information needed for the program initiation milestone review. OIPTs should meet, thereafter, as necessary over the life of the program. The OIPT leader should act to resolve issues when requested by any member of the OIPT, or when so directed by the Milestone Decision Authority. The goal is to resolve as many issues and concerns at the lowest level possible, and to expeditiously escalate issues that need resolution at a higher level. The OIPT should bring only the highest-level issues to the Milestone Decision Authority for decision.

The OIPT should normally convene 2 weeks before a planned decision point. It should assess the information and recommendations that the Milestone Decision Authority will receive. It should also assess family-of-system or system-of-system capabilities within and between functional portfolios (or areas) in support of integrated architectures developed by the Joint Staff in collaboration with the OSD, USAF (as DoD Space Milestone Decision Authority), and the DoD Components. If the program includes a pilot project, such as Total Ownership Cost Reduction, the Program Manager should report the status of the project to the OIPT. The OIPT should then assess progress against stated goals. The Program Manager's briefing to the OIPT should address interoperability and supportability (including spectrum supportability) with other systems, anti-tamper provisions, and indicate whether those requirements will be satisfied by the acquisition strategy under review. If the program is part of a family-of-systems architecture, the Program Manager should brief the OIPT in that context. If the architecture includes less than Acquisition Category I programs that are key to achieving the expected operational capability, the Program Manager should also discuss the status of and dependence on those programs. The OIPT should review the programmatic risk issues of cost, schedule, and performance. The OIPT leader should recommend to the Milestone Decision Authority whether the anticipated review should go forward as planned.

For Acquisition Category ID decision points, the OIPT leader will provide the Defense Acquisition Board chair, co-chair, principals, and advisors with an integrated assessment using information gathered through the IPPD process. The OIPT assessment should focus on core acquisition management issues and should consider independent assessments, including technology readiness assessments, which the OIPT members normally prepare. These assessments typically occur in context of the OIPT review, and should be reflected in the OIPT leader's report. There should be no surprises at this point-all team members should work issues in real time and should be knowledgeable of their OIPT leader's assessment. OIPT and other staff members should minimize requirements for the program manager to provide pre-briefs independent of the OIPT process.

PHYSICAL CONFIGURATION AUDIT (PCA)

The Defense Acquisition Guidebook describes the PCA as follows:[22]

The PCA is conducted around the time of the full rate production decision. The PCA examines the actual configuration of an item being produced. It verifies that the related design documentation matches the item as specified in the contract. In addition to the standard practice of assuring product verification, the PCA confirms that the manufacturing processes, quality control system, measurement and test equipment, and training are adequately planned, tracked, and controlled. The PCA validates many of the supporting processes used by the contractor in the production of the item and verifies other elements of the item that may have been impacted/redesigned after completion of the System Verification Review (SVR). A PCA is normally conducted when the government plans to control the detail design of the item it is acquiring via the Technical Data Package. When the government does not plan to exercise such control or purchase the item's Technical Data Package (e.g., performance based procurement) the contractor should conduct an internal PCA to define the starting point for controlling the detail design of the item and establishing a product baseline. The PCA is complete when the design and manufacturing documentation match the item as specified in the contract. If the PCA was not conducted prior to the full rate production decision, it should be performed as soon as production systems are available.

POST CRITICAL DESIGN REVIEW ASSESSMENT (PCDRA)

Department of Defense Instruction 5000.02 describes the PCDRA as follows:[23]

[22]Defense Acquisition Guidebook, Section 4.3.4.4.3, Physical Configuration Audit (PCA). Available online at http://akss.dau.mil/dag/DoD5000.asp?view=document&rf=GuideBook\IG_c4.3.4.4.3.asp. Last accessed May 4, 2009.

[23]Department of Defense Instruction 5000.02, Operation of the Defense Acquisition System, December 8, 2008.

The MDA shall conduct a formal program assessment following system-level CDR. The system-level CDR provides an opportunity to assess design maturity as evidenced by measures such as: successful completion of subsystem CDRs; the percentage of hardware and software product build-to specifications and drawings completed and under configuration management; planned corrective actions to hardware/software deficiencies; adequate developmental testing; an assessment of environment, safety and occupational health risks; a completed failure modes and effects analysis; the identification of key system characteristics; the maturity of critical manufacturing processes; and an estimate of system reliability based on demonstrated reliability rates.

1. The PM shall provide a Post-CDR Report to the MDA that provides an overall assessment of design maturity and a summary of the system-level CDR results which shall include, but not be limited to:

a. The names, organizations, and areas of expertise of independent subject matter expert participants and CDR chair;

b. A description of the product baseline for the system and the percentage of build-to packages completed for this baseline;

c. A summary of the issues and actions identified at the review together with their closure plans;

d. An assessment of risk by the participants against the exit criteria for the EMD Phase; and

e. Identification of those issues/risks that could result in a breach to the program baseline or substantively impact cost, schedule, or performance.

2. The MDA shall review the Post-CDR Report and the PM's resolution/ mitigation plans and determine whether additional action is necessary to satisfy EMD Phase exit criteria and to achieve the program outcomes specified in the APB. The results of the MDA's Post-CDR Assessment shall be documented in an ADM.

3. Successful completion of the Post-CDR Assessment ends Integrated System Design and continues the EMD Phase into System Capability and Manufacturing Process Demonstration.

PRELIMINARY DESIGN REVIEW (PDR)

The Defense Acquisition Guidebook describes the PDR as follows:[24]

---

[24]Defense Acquisition Guidebook, Section 4.3.3.4.4, Preliminary Design Review (PDR). Available online at https://akss.dau.mil/dag/GuideBook/IG_c4.3.3.4.4.asp. Last accessed May 4, 2009.

The PDR is a multi-disciplined technical review to ensure that the system under review can proceed into detailed design, and can meet the stated performance requirements within cost (program budget), schedule (program schedule), risk, and other system constraints. Generally, this review assesses the system preliminary design as captured in performance specifications for each configuration item in the system (allocated baseline), and ensures that each function in the functional baseline has been allocated to one or more system configuration items. Configuration items may consist of hardware and software elements and include such items as airframes, avionics, weapons, crew systems, engines, trainers/training, etc.

PRODUCTION READINESS REVIEW (PRR)

The Defense Acquisition Guidebook describes the PRR as follows:[25]

The PRR examines a program to determine if the design is ready for production and if the producer has accomplished adequate production planning. The review examines risk; it determines if production or production preparations incur unacceptable risks that might breach thresholds of schedule, performance, cost, or other established criteria. The review evaluates the full, production-configured system to determine if it correctly and completely implements all system requirements. The review determines whether the traceability of final system requirements to the final production system is maintained.

PROGRAM EXECUTIVE OFFICER SUFFICIENCY REVIEW (PEO/SR)

The PEO/SR was described to the committee as follows:[26]

The purpose of the AAC PEO Program Sufficiency Review Process is three-fold. First, to provide acquisition teams a robust support structure to develop/refine acquisition strategies. The second purpose is to effectively employ senior acquisition leaders' experience with independent/peer reviews of high-interest acquisition plans, program technical status along with associated cost and schedule. Finally, the review assists the program team to design realistic/high confidence program plans and to be able to understand, agree with and fully explain the risks of the program. Sufficiency Reviews are the final step in an integrated assessment of technical and programmatic elements of the program construct.

---

[25]Defense Acquisition Guidebook, Section 4.3.3.9.3, Production Readiness Review (PRR). Available online at https://akss.dau.mil/dag/GuideBook/IG_c4.3.3.9.3.asp. Last accessed May 4, 2009.

[26]Judy A. Stokley, SES, USAF AFMC AAC/CA e-mail to Jim Garcia, January 9, 2009.

PROGRAM SUPPORT REVIEW (PSR)

Department of Defense Instruction 5000.02 describes the PSR as follows:[27]

> PSRs are a means to inform an MDA and Program Office of the status of technical planning and management processes by identifying cost, schedule, and performance risk and recommendations to mitigate those risks. PSRs shall be conducted by cross-functional and cross-organizational teams appropriate to the program and situation. PSRs for ACAT ID and IAM programs shall be planned by the Director, Systems and Software Engineering (SSE) to support OIPT program reviews, at other times as directed by the USD(AT&L), and in response to requests from PMs.

SYSTEM DESIGN REVIEW (SDR)

On the Defense Acquisition University's Acquisition Community Connection Web site, the SDR is described as follows:[28]

> This review is conducted to evaluate the optimization, correlation, completeness, and risks associated with the allocated technical requirements. A review of the system engineering process that produced the allocated technical requirements and of the engineering planning for the next phase of effort should also be reviewed. Basic manufacturing considerations should be reviewed and planning for production engineering in subsequent phases should be addressed. This review should be conducted when the system definition effort has proceeded to the point where system characteristics are defined and the configuration items are identified.

SYSTEM FUNCTIONAL REVIEW (SFR)

The Defense Acquisition Guidebook describes the SFR as follows:[29]

> The SFR is a multi-disciplined technical review to ensure that the system under review can proceed into preliminary design, and that all system requirements and functional performance requirements derived from the Capability Development Document are defined and are consistent with cost (program budget), schedule (program schedule), risk, and other system constraints. Generally this review assesses the system functional requirements as captured in system specifications (functional baseline), and ensures that all required system performance is fully

---

[27]Department of Defense Instruction 5000.02, Operation of the Defense Acquisition System, December 8, 2008.

[28]Available online at https://acc.dau.mil/CommunityBrowser.aspx?id=50742&lang=en-US. Last accessed May 5, 2009.

[29]Defense Acquisition Guidebook, Section 4.3.3.4.3, System Functional Review (SFR). Available online at https://akss.dau.mil/dag/GuideBook/IG_c4.3.3.4.3.asp. Last accessed May 4, 2009.

decomposed and defined in the functional baseline. System performance may be decomposed and traced to lower-level subsystem functionality that may define hardware and software requirements. The SFR determines whether the systems functional definition is fully decomposed to a low level, and whether the IPT is prepared to start preliminary design.

SYSTEM REQUIREMENTS REVIEW (SRR)

The Defense Acquisition Guidebook describes the SRR as follows:[30]

The SRR is conducted to ascertain progress in defining system technical requirements. This review determines the direction and progress of the systems engineering effort and the degree of convergence upon a balanced and complete configuration. It is normally held during Technology Development, but may be repeated after the start of System Development and Demonstration to clarify the contractor's understanding of redefined or new user requirements.

The SRR is a multi-disciplined technical review to ensure that the system under review can proceed into the System Development and Demonstration phase, and that all system requirements and performance requirements derived from the Initial Capabilities Document or draft Capability Development Document are defined and are consistent with cost (program budget), schedule (program schedule), risk, and other system constraints. Generally this review assesses the system requirements as captured in the system specification, and ensures that the system requirements are consistent with the preferred system solution as well as available technologies resulting from the Technology Development phase. Of critical importance to this review is an understanding of the program technical risk inherent in the system specification and in the System Development and Demonstration Phase Systems Engineering Plan. Determining an acceptable level of risk is key to a successful review.

SYSTEM VERIFICATION REVIEW (SVR)

The Defense Acquisition Guidebook describes the SVR as follows:[31]

The SVR is a multi-disciplined product and process assessment to ensure that the system under review can proceed into Low-Rate Initial Production and Full-Rate Production within cost (program budget), schedule (program schedule), risk, and other system constraints. Generally this review is an audit trail from the Critical Design Review. It assesses the system final product, as evidenced in

[30]Defense Acquisition Guidebook, Section 4.3.2.4.1, System Requirements Review (SRR). Available online at https://akss.dau.mil/dag/GuideBook/IG_c4.3.2.4.asp#43241. Last accessed May 4, 2009.

[31]Defense Acquisition Guidebook, Section 4.3.3.9.2. System Verification Review (SVR), https://akss.dau.mil/dag/GuideBook/IG_c4.3.3.9.2.asp. Last accessed May 4, 2009.

its production configuration, and determines if it meets the functional requirements (derived from the Capability Development Document and draft Capability Production Document) documented in the Functional, Allocated, and Product Baselines. The SVR establishes and verifies final product performance. It provides inputs to the Capability Production Document. The SVR is often conducted concurrently with the Production Readiness Review. A Functional Configuration Audit may also be conducted concurrently with the SVR, if desired.

TECHNOLOGY READINESS ASSESSMENT (TRA)

The Defense Acquisition Guidebook describes the TRA as follows:[32]

The TRA is a systematic, metrics-based process that assesses the maturity of Critical Technology Elements. The TRA should be conducted concurrently with other Technical Reviews, specifically the Alternative Systems Review, System Requirements Review, or the Production Readiness Review. If a platform or system depends on specific technologies to meet system operational threshold requirements in development, production, and operation, and if the technology or its application is either new or novel, then that technology is considered a Critical Technology Element. The TRA should not be considered a risk assessment, but it should be viewed as a tool for assessing program risk and the adequacy of technology maturation planning. The TRA scores the current readiness level of selected system elements, using defined Technology Readiness Levels. The TRA highlights critical technologies and other potential technology risk areas that require program manager attention. The TRA essentially "draws a line in the sand" on the day of the event for making an assessment of technology readiness for critical technologies integrated at some elemental level. If the system does not meet pre-defined Technology Readiness Level scores, then a Critical Technology Element maturation plan is identified. This plan explains in detail how the Technology Readiness Level will be reached prior to the next milestone decision date or relevant decision point.

TEST READINESS REVIEW (TRR)

The Defense Acquisition Guidebook describes the TRR as follows:[33]

The TRR is a multi-disciplined technical review to ensure that the subsystem or system under review is ready to proceed into formal test. The TRR assesses test objectives, test methods and procedures, scope of tests, and safety and confirms that required test resources have been properly identified and coordinated

---

[32]Defense Acquisition Guidebook, Section 4.3.2.4.3. Technology Readiness Assessment (TRA). Available online at https://akss.dau.mil/dag/DoD5000.asp?view=document&rf=GuideBook\IG_c4.3.3.9.4.asp. Last accessed May 4, 2009.

[33]Defense Acquisition Guidebook, Section 4.3.3.9.1. Test Readiness Review (TRR). Available online at https://akss.dau.mil/dag/GuideBook/IG_c4.3.3.9.asp#43391. Last accessed May 4, 2009.

to support planned tests. The TRR verifies the traceability of planned tests to program requirements and user needs. The TRR determines the completeness of test procedures and their compliance with test plans and descriptions. The TRR assesses the system under review for development maturity, cost/schedule effectiveness, and risk to determine readiness to proceed to formal testing. In addition to adequate planning and management, to be effective the program manager should follow-up [sic] with the outcomes of the TRR.